MULTIPLE SCALES OF SUSPEN DYNAMICS IN A COMPLEX GEC

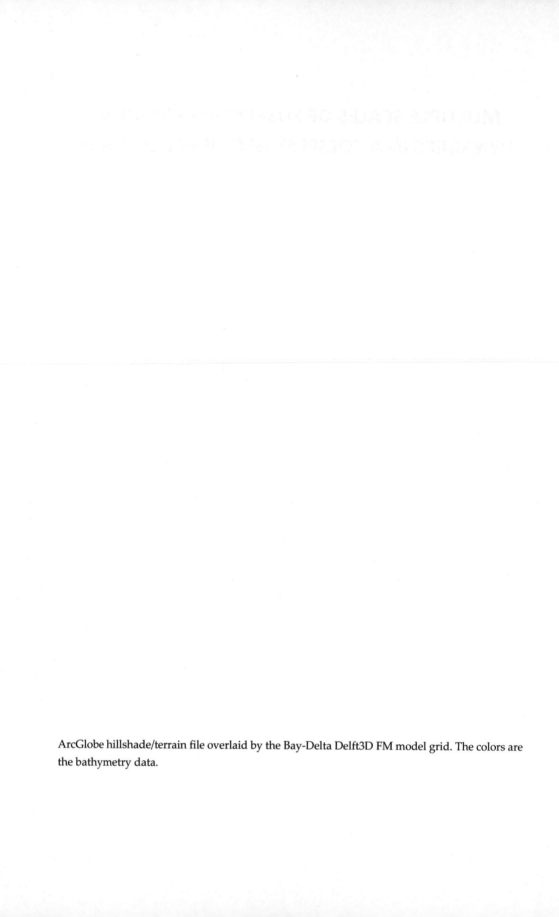

ArcGlobe hillshade/terrain file overlaid by the Bay-Delta Delft3D FM model grid. The colors are the bathymetry data.

MULTIPLE SCALES OF SUSPENDED SEDIMENT DYNAMICS IN A COMPLEX GEOMETRY ESTUARY

DISSERTATION

Submitted in fulfillment of the requirements of
the Board for Doctorates of Delft University of Technology
and
of the Academic Board of the UNESCO-IHE
Institute for Water Education
for
the Degree of DOCTOR
to be defended in public on
Tuesday, 12 April 2016, 15:00 hours
In Delft, the Netherlands

by

Fernanda MINIKOWSKI ACHETE

Master of Science in Coastal and Marine Engineering and Management
Technology University Delft
Norwegian Technology University
Technology University Catalunia

born in Rio de Janeiro, Brazil

This dissertation has been approved by the
promotor: Prof. dr. ir. J.A. Roelvink and
copromotor: Dr. M. van der Wegen

Composition of the Doctoral Committee:

Chairman	Rector Magnificus, Delft University of Technology
Vice-Chairman	Rector UNESCO-IHE
Prof. dr. ir. J.A. Roelvink	UNESCO-IHE /Deltares/ Delft University of Technology, promotor
Dr. ir. M. van der Wegen	UNESCO-IHE/ Deltares, copromotor

Other member:

Dr. B. Jaffe	U.S. Geological Survey Pacific Science Center

Independent members:

Prof. dr. ir. J.C. Winterwerp	Delft University of Technology
Prof. dr. ir. H.H.G. Savenije	Delft University of Technology
Prof. dr. M.E. McClain	UNESCO-IHE / Delft University of Technology
Dr. ir. A.J. Kettner	University of Colorado Boulder, INSTAAR
Prof. dr. D.P. Solomatine	UNESCO-IHE / Delft University of Technology, reserve member

This research was conducted under the auspices of the Graduate School for Socio-Economic and
Natural Sciences of the Environment (SENSE)

CRC Press/Balkema is an imprint of the Taylor & Francis Group, an informa business

Published by:
CRC Press/Balkema
PO Box 11320, 2301 EH Leiden, The Netherlands
Pub.NL@taylorandfrancis.com
www.crcpress.com – www.taylorandfrancis.com
ISBN 978-1-138-.02976-7

"Great things are not done by impulse,

but by a series of small things brought together."

— Vincent Van Gogh

Acknowledgments

First and foremost, I would like to express my sincere gratitude to my supervisory team Prof. Dano Roelvink, Dr. Mick van der Wegen and Dr. Bruce Jaffe for their insightful supervision and guidance enabling me to become a researcher. Dano, I want to thank you for the milestone meetings with accurate remarks, pointing multiple paths and trusting in my work. You are an example to be followed with sharp eyes for science and sharp ears for music. Mick, your passion for your work inspired me to spread the word of my own research. Thank you for the infinite patience reading my documents and teaching me the power of a good talk. Bruce, even far away you always had the right questions to pose. I am grateful for you taking the time in your busy agenda to accommodate my work and showing me how to look at the Bay area and beyond.

I gratefully acknowledge the funding sources of my Ph.D. from USGS/CALFED, the Dutch Ministry of Infrastructure and Environment via fellowship MoU with UIHE., and Capes. This work was embedded in the CASCaDE project (Computational Assessments of Scenarios of Change for Delta Ecosystem, contribution 61), and accounted with the collaboration of several colleagues that were crucial for the completion of this work. Special thanks to Lisa Lucas, Noah Knowles and Rose Martyr for the fruitful discussions and invaluable support during my periods in California. My deepest gratitude goes to David Schoellhamer, Scott Wright, Gregory Shellenbarger and Guy Gelfenbaum for sharing their knowledge and providing me with many questions to be answered which enriching this work. I would also like to thanks, Theresa Fregoso, Amy Foxgrover and Tara Morgan for the precious support providing me data and beautiful figures and Tara Schraga for taking me into the boat and then into your life. Herman Kernkamp, Arthur van Dam, Sander van der Pijl, Arjen Markus, and Michel de Jeuken, thank you for the lengthy discussions about the model equations and explained me every new feature.

My sincere gratitude goes to Marc Sas, from IMDC, for hosting me during most of my Ph.D. time, providing me not just a desk but an inspiring working environment and many colleagues, Gijsbert van Holland, Julien da Silva, Marion Coquet and Rohit Kulkarni.

My time in the CSEPD at UNESCO-IHE would not be the same without the support of my colleagues Rosh Ranasinghe, Ali Dastgheib, Miréia Lopez, Guo Leicheng, Leo Sembiring, Trang Duong, Johan Reyns, Abdi, and Janaka. Thank you for the technical and the not-so-technical discussions.

Gerald Corzo, thank you for introducing me to the Surf Sara environment, it saved me precious running hours (months). Afonso Paiva, you always going be my first professor. Susana Vinzon and Eduardo Siegle, I appreciate all the support that you gave me in Delft and in Brazil.

During my time here I received a couple of special gifts, my dear friends. Gabriel Kogan, your short time here were full of special moments that propagates all the way to São Paulo. Aline Kaji and AmauryCamarena, thank you for making me feel home and ours long discussions about basically everything. Angelo Vallozzi, your kindness, and a beautiful smile fill my days with joy. Patricia, you inspired me as a professional and as a friend. Aki, yours never ending curiosity is a driver always to look further. Juan Carlos Chacon, your hug, and laugh are like the sun. Anna Potysz, our connection, and understanding transformed me. Fernanda Braga, thanks for making my life simple and crazy. I am grateful for the beautiful friends that I made here: Vero, Yared, Jessica, Chris, Angelica, Marcello, Giorgio, Sophie, Laura, Thaine, Mohan, thank you for all the support.

To my dearest friends in Brazil, the 2 years became almost 7, but besides the distance our bounds just grew stronger, and the visits to Brazil completely recharged my batteries Pops our friendship is precious.

To the ones that were there since ever. Marlene and Carlos, you are my source of inspiration, your care and support made my achievements possible. Thank you for giving me wings, showing that the world is there to be conquered and giving me the 2 most precious gifts of my life Janaína and Pedro. Jana, my beloved sister, thank you for always being by my side, teaching me how to argue, and shown me that pure love can exists bringing to us Marina. Pedro, you showed me that we can always have more energy. And to my grammas that still give me the honor enjoy their company and knowledge. I love you.

At last but not least, I want to express all my love to Bruno, for being the incredible and understanding person that you are and giving me all the emotional and technical support. For bringing into my life Silvia, Livia, Xavier, and Luca; for all our endless walks to discuss my topic, for being supportive with the ups and downs that I went through along the Ph.D. process. We grew together, and I am really proud of it.

Para meus queridos amigos que ficaram no Brasil, 2 anos tornaram-se quase 7, mas além da distância nosso relacionament só fortificou, as visitas ao Brasil tinham o incrível poder de recarregar minhas baterias com muitas risadas e amor. Pops nossa amizade é preciosa.

Para os que estavam lá desde sempre. Marlene e Carlos, vocês são minha cosntante fonte de inspiração, seus cuidados e apoio fizeram com que minhas realizações se tornassem possíveis. Obrigado por me dar asas, mostrando que o mundo está lá para ser conquistado e acima de tudo me dando os 2 presentes mais preciosos da minha vida Janaína e Pedro. Jana, minha amada irmã, obrigado por estar sempre ao meu lado, me ensinando como argumentar (discutir), e me mostrar o que o amor puro pode existir trazendo para nós Marina. Pedro, você me provou que podemos sempre ter mais energia. E às minhas avós que ainda me dão a honra de poder compartilhar sua compania e sabedoria. Eu amo vocês.

Por último mas não menos importante, quero expressar todo meu amor para Bruno, por ser essa pessoa incrível, por sua compreensão e me dando todo o apoio emocional e técnico nos momentos que mais precisava. Por trazer à minha vida Silvia, Livia, Xavier, e Luca; por todas as nossas intermináveis caminhadas para discutir o meu tema, por ser solidário com todos os altos e baixos que eu passei ao longo do processo do Ph.D.. Nós crescemos juntos, e sou muito orgulhosa disso.

Summary

Many estuaries are located in urbanized, highly engineered environments. At the same time, they host valuable ecosystems and natural resources. These ecosystems rely on the maintenance of habitat conditions which are constantly changing due to impacts like sea level rise, reservoir operations, and other civil works. As for many estuarine systems, cohesive sediment plays an important role. It is important not only because of its physical behavior but also due to its link with ecology. An important ecological link is the suspended sediment concentration translated in turbidity levels and sediment budget.

The main objective of this Ph.D. work is to investigate turbidity levels and sediment budget variability at a variety of spatial and temporal scales and the way these scales interact. We use the San Francisco Bay-Delta system as study case because it has been well measured over a long time period, it has a complex geometry including embayments and many channels, and it provides habitat, nursery and trophic support for several endemic species and has been subject to extensive human interference.

This study has been developed using a finite volume, process-based model, D-FLOW Flexible Mesh, which allows coupling complex geometry river, estuary and coastal systems in the same grid. Delft3D FM also allows for direct coupling with ecological model. This thesis shows that with simple model settings, for example of one single mud fraction and simple bottom sediment distribution it is possible to have a robust sediment model, which reproduces 90% of yearly sediment budget comparing to data derived budget.

This study explores the sediment dynamics variability and sediment budget for different scales. It compares the dynamics of a large scale system like the Sacramento-San Joaquin Delta, O(100km; days-weeks), to the much smaller system of Alviso Slough, O(10km; hours). Based on the prevailing sediment dynamics we can subdivide these systems into event-driven and a tide-driven estuary (table 1). The scenarios simulations show possible changes in the estuaries sediment dynamics. In the event-driven estuary (Delta) reducing the river sediment input impacts the entire system dynamics. In the tide-driven estuary (Alviso Slough) the opening of ponds abruptly changes the tidal prism and tidal propagation leading to significant erosion and deposition areas in the Slough.

Table 1: Comparison between main characteristics of a tidal and an event-driven estuary.

	Event-driven estuary (Delta)	Tide-driven Estuary (Alviso)
Main Sediment Forcing	River Discharge	Tides
SSC timescale	Days - weeks	Hours
Morphodynamic adaptation time scale	Weeks - months	Years
Boundary	Landward	Seaward
Main sediment calibration parameter	Fall Velocity	Erosion Coefficient
The main sediment transport direction	Unidirectional - watershed towards the Bay	Bidirectional

This thesis also makes advances in connecting to other science fields and develops a managerial tool that is able to support the decision-making process. The calibrated and validated model is a powerful tool for managers.

Samenvatting

Veel estuaria zijn gelegen in een verstedelijkte, door mensen bepaalde omgeving. Tegelijkertijd herbergen ze waardevolle ecosystemen en natuurlijke hulpbronnen. Deze ecosystemen zijn afhankelijk van een goed onderhoud van habitat omstandigheden welke voortdurend veranderen als gevolg van zeespiegel stijging, dam reservoir management en andere civiele werken. In veel estuariene systemen speelt cohesief sediment een belangrijke rol in de gezondheid van het ecosysteem. Een belangrijke ecologische link is de suspensieve sediment concentratie vertaald naar troebelheid niveaus en sediment budget.

De belangrijkste doelstelling van dit proefschrift is om variabiliteit van troebelheid en sediment budget op verschillende ruimtelijke schalen en tijd schalen te onderzoeken met inbegrip van de interactie tussen deze schalen. We nemen het San Francisco Bay-Delta systeem als uitgangspunt, omdat het systeem goed is bemeten over een langere periode, het systeem een complexe geometrie heeft inclusief kleinere baaien en veel kanalen, het een habitat en voedingsbodem vormt voor verschillende endemische soorten en het onderhevig is geweest aan significant menselijk ingrijpen.

Deze studie maakt gebruik van een proces-gebaseerd model op basis van eindige volumes, D-FLOW Flexible Mesh (FM). Deze software staat koppeling toe van grids in rivieren, estuaria en kustsystemen met een complexe geometrie. Delft3D FM staat ook een directe koppeling met ecologische modellen toe. Dit proefschrift laat zien dat met eenvoudige modelinstellingen (bijvoorbeeld één slib fractie en een simpele bodem sediment distributie) het mogelijk is om een robuust sediment model te maken dat 90% van het jaarlijkse gemeten sediment budget goed voorspelt.

Deze studie verkent de variatie in sediment dynamica en sediment budgetten voor verschillende schalen. De studie vergelijkt de dynamica van een grootschalig systeem zoals de Sacramento-San Joaquin Delta, O(100km; dagen-weken), met het veel kleinere systeem van Alviso Slough, O(10km; uur). Gebaseerd op de heersende sediment dynamica kunnen we deze systemen onderverdelen in een 'event' gedreven estuarium en een getij gedreven estuarium (tabel 1). De scenario runs laten mogelijke veranderingen zien in de estuariene sediment dynamica. In het 'event' gedreven estuarium (de Delta) heeft een afname van de rivier sediment toevoer impact op het hele domein. In het getij gedreven systeem (Alviso slough) verandert de opening van de zout pannen het getij prisma en de getij voortplanting abrupt, wat leidt tot significante erosie en depositie in bepaalde gebieden van de Slough.

Tabel 1: Vergelijking tussen de belangrijkste kenmerken van een getijde en een 'event' gedreven estuarium.

	Event gedreven estuarium (Delta)	Getij gedreven estuary (Alviso slough)
Voornaamste sediment forcering	Rivier debiet	Getij
SSC tijdschaal	Dagen - weken	Uren
Moprhodynamische adaptatie tijdschaal	Weken - maanden	Jaren

	Event gedreven estuarium (Delta)	Getij gedreven estuary (Alviso slough)
Rand	Landwaarts	Zeewaarts
Voornaamste sediment calibratie parameter	Valsnelheid	Erosie coefficient
Voornaamste sediment transport richting	Eén richting - zeewaarts	Twee richtingen

Deze studie verbindt ook andere wetenschappelijke disciplines en ontwikkelt een beheersinstrument welke gebruikt kan worden om het besluitvormingsproces te ondersteunen. Het gekalibreerde en gevalideerde model is een krachtig instrument voor bestuurders.

Resumo

Muitos estuários estão localizados em áreas altamente urbanizadas, e industrializado. Ao mesmo tempo, há valiosos ecossistemas e os recursos naturais. Estes ecossistemas dependem da manutenção de habitat para sobreviverem, poré essas condições mudam constantemente devido a impactos como o aumento do nível do mar, operações de reservatórios e outras obras civis. Como para muitos sistemas estuarinos, sedimentos coesivos desempenham um papel importante, não só por seu comportamento físico, mas também devido à sua relação com a ecologia. Um parâmetro ecológico importante é a concentração de sedimentos em suspensão traduzido em níveis de turbidez e o balanço de sedimentos.

O principal objetivo deste Ph.D. é investigar os níveis de turbidez e variabilidade do balanço de sedimentos em uma diversas escalas espaciais e temporais, e modo como essas escalas interagem. San Francisco Bay-Delta é utilizada como estudo de caso, porque tem sido um extenso banco de dados, tem uma geometria complexa, incluindo baías e canais, é habitat, berçário e suporte trófico para várias espécies endêmicas e tem sido palco de extensa interferência humana.

Este estudo foi desenvolvido utilizando um modelo baseado em processos de volumes finitos, D-FLOW malha flexível (Delft3D FM), que permite englobar o rio, estuário e sistema costeiro na mesma grade. Delft3D FM também permite o acoplamento com modelo ecológico. Esta tese mostra que, com configurações de modelo simples, por exemplo, com uma fracção de lama e distribuição sedimento de fundo simples, é possível ter um modelo de sedimentos robusto, reproduzindo 90% do balanço de sedimentos.

Este estudo explora a variabilidade de dinâmica e balanço sedimentar para diferentes escalas. Ele compara a dinâmica de um sistema de larga escala como o Sacramento-San Joaquin Delta, O (100km; dias-semanas), com um muito menor o Alviso Slough, O (10km; horas). Com base na dinâmica sedimentar predominante, podemos subdividir esses sistemas em forçados por eventos e forçados por marés (tabela 1). No estuário forçado por eventos (Delta) a redução de sedimento na fronteira fluvial altera toda a dinâmica do sistema. No estuário forçado por marés (Alviso Slough) a abertura abrupta de lagoas modifica o prisma de maré, consequentemente a propagação da maré levando a significativos processos de erosão e deposição no leito.

Tabela 1: Comparação entre as principais caracterísitcas entre estuários forçados por eventos e por marés.

	Estuário forçados por eventos (Delta)	Estuário Forçado por marés (Alviso)
Principal forçante para a dinâmica de sedimentos	Descarga de rio	Marés
Escala de tempo de sedimento em suspenção	Dias-semanas	Horas
Escala de tempo de adaptação morfológica	Semanas-meses	Anos
Fronteiras	Terrestre	Marítima
Principal parâmetro de calibração	Velocidade de queda	Coeficiente de erosão

Direção principal de transporte de sedimento	Unidirecional- da bacia de drenagem para a Baía	Bi-direcional

Esta tese também faz avanços na conexão com outros campos da ciência e desenvolve uma ferramenta de gerenciamento capaz de auxiliar o processo de tomada de decisão. O modelo calibrado e validado é uma poderosa ferramenta para gestores

Contents

1

GENERAL INTRODUCTION

1.1 Background

Estuaries are valuable ecosystems for flora, fauna, and human beings. Society, as we know today, has a part of its history related to these environments where prosperous civilization developed (Day et al., 2007). Estuaries, more specifically deltas and flood plains, provide rich soil for agriculture, due to fine sediments deposited on the banks and fishery stocks. They provide a sheltered area which is favorable to host port facilities, to link coastal to inland navigation, and consequently industry settlement. Industry, in most cases, exploits natural resources and may contribute to the contamination of water and soil (mostly fine sediment).

Estuaries provide a complex system of barotropic and baroclinic currents and sediment transport due to tidal incursion and freshwater discharge that can present highly seasonal variation (Dyer, 1986). Considering abiotic characteristics, mixing of fresh and salt water occurs in estuaries generating a salinity gradient zone. This gradient creates a wide range of habitat conditions promoting rich flora and fauna, as well as biodiversity including endemic species.

As for many estuarine systems, cohesive sediment plays an important role. It is important not only because of its physical behavior but also due to the link with estuarine health and ecology. A number of reasons can be mentioned:

a) Fine sediment is the most accountable for vegetation colonization since it stabilizes the substrate and retains nutrients. It plays an important role in colonization of sub-aquatic vegetation and marsh maintenance as well as the ability of this vegetation to cope with sea level rise (Morris et al., 2002).

b) It regulates phytoplankton productivity that is the aquatic food web base (Burkholder, 1992). On one hand, it increases the water turbidity, decreasing the optic depth. On the other hand, carries nutrients essential for phytoplankton development so enhancing the production;

c) Cohesive sediments are responsible for river bank fertilization. During flood conditions the fine sediment overtops the banks and deposits on the riverside; due to associated high organic matter concentration the sediment works as a fertilizer;

d) Organic (e.g. PCB's) and inorganic (e.g. heavy metal) contaminants can easily adhere to fine sediment increasing the contaminants' residence time in the system so facilitating the entrance in the food web (Winterwerp and Van Kesteren, 2004);

e) Fine sediment tends to accumulate in still water such as bays and inside harbors, increasing maintenance dredging costs to ensure port accessibility.

Taking into account the aforementioned factors it is important to understand estuarine systems and to develop the capacity of forecasting sediment path, system turbidity (which is proportional to turbidity) and sediment budget.

1.1.1 Hydrodynamics

There are many factors that influence sediment dynamic and availability as tides, river discharge, wind, waves, dredging, and superficial flow in adjacent land. In this section, we are going introduce the main hydrodynamic processes that govern the sediment dynamics

1.1.1.1 Tides

In a simplistic way, tides are periodic sea surface variation forced by the Earth-Moon-Sun gravitational attraction. Tides can be seen as long waves propagating in the ocean basin and if we consider no friction, no rotation, no inertia and no obstacles we have a rising and falling cycle every 12 hours. However, this is not the case, especially considering propagation inside estuaries.

The tidal wave excursion landwards defines the estuary extension. The tides while penetrating in estuaries suffer distortion, attenuation or amplification (Friedrichs and Aubrey, 1988; Savenije, 2001). These processes modify the duration of ebb and flood tides creating asymmetries, in estuaries worldwide both ebb or flood asymmetries can be observed.

This means that there is an inequality between ebb and flood current, this difference will create a residual sediment transport defining whether there is an import or export of sediment. The freshwater discharge also influences in this duration (Ridderinkhof et al., 2000; Talke and Stacey, 2003).

Tides also transport salt landwards what creates an environment with salinity gradient and so a density gradient zone that is going to be described afterward. This brackish environment is important to many aquatic species that nurse and feed in such a system.

1.1.1.2 Freshwater flow

This work considers freshwater input exclusively by the rivers to simplify the analysis and it is the source which there is available data. The fresh water flow is the landward boundary of an estuary, which provides fresh water and is the main source of sediment to estuaries. The watershed drainage carries sediment from upstream that later is going to feed estuary and coastal region.

In many estuaries, the freshwater discharge accounts for a big portion of the system variability. It is usual to observe an annual cycle of dry and wet season. Freshwater discharge is going to modulate the salinity gradient, the estuarine circulation, water quality (including turbidity), productivity and abundance of species.

1.1.1.3 Wind and wind-waves

The wind is an important energy source for the system. In estuaries, the wind may generate waves that stir sediment from the bottom. The stirring process is due to waves orbital velocities that can feel the bottom in shallow water. Orbital velocity is proportional to the wave height, so the wind and wind-wave impacts can be better seen in shallow regions since in estuaries the fetch is limited and so the wave heights. In some cases, it is possible to observe the penetration of ocean waves inside the estuaries. In these cases, the stirring process can be felt in deeper waters (Talke and Stacey, 2003).

The importance of stirring sediment from mud-flats and shoals can be enhanced by the tidal excursion, so the waves stir the sediment while the tidal currents transport it. This process will be further discussed below in the sediment dynamics section.

1.1.2 Sediment dynamics

In estuaries, most of the sediment input are delivered in suspension by rivers, resulting in an hourly sediment load that is dependent on the river discharge. River discharge and consequently SSC may vary on weekly to yearly time scales. When considering sea level rise, the time scale is increased to centuries.

The sediment dynamics is closely linked to the hydrodynamic processes described above. However, first is important to identify if the sediment is cohesive or not. Cohesive sediment (mud) has completely different dynamics than non-cohesive sediment (sand). Cohesive sediment dynamics, for example, are affected by flocculation, lower fall velocity, and consolidation and higher stress for erosion (Winterwerp and Van Kesteren, 2004).

The difference arises from the electrochemical interactions between clay particles, so the cohesiveness of sediment depends on the clay content. Laboratory experiments show that sediment becomes cohesive when the clay content is over 3-5% (Van Ledden et al., 2004). Though, the clay percentage necessary to make the sediment cohesive may increase with larger sand grains (Le Hir et al., 2011).

Currents and waves loose energy in the bottom boundary layer, this energy dissipation is translated in shear stress at the bed (Grant and Madsen, 1979). Sedimentation takes place while the shear stress is below a critical value; above the critical value, erosion takes place. The critical shear stress depends on the bottom composition; cohesiveness can increase the critical erosion shear stress in 2-5 times (Van Ledden et al., 2004). This means that the velocities to erode the sediment layer should be higher even though the sediment is finer.

During flood tide, one observes a peak of sediment concentration that it is advected landward. During the slack water, the sediment has time enough to settle and partially consolidate and is not re-suspended during ebb tide (Postma, 1961). This is the main sediment transport forcing for tide-driven estuaries. This transport scenario changes in stormy periods becoming ebb dominated. For event-driven estuaries, during wet periods the peak river events with high SSC dominate the sediment transport in the estuary (Coco et al., 2007; Dyer et al., 2000; Ralston and Stacey, 2007), overcoming the tidal sediment transport.

1.1.2.1 Sediment Budget

In this section, we are going to discuss the sediment origin (source) and destination (sink). Looking from a broader perspective the highland is the source of sediment, the rivers transport it, the estuaries is the interface between the land and the sea, and the ocean is the final destination.

Nevertheless, many processes may enhance or decrease sediment supply, disrupting the sediment budget.

Starting from a worldwide perspective,(Syvitski and Kettner, 2011), state that the human impact on sediment production dates from 3000 years ago accelerating in the last 1000 due to engineering developments. In the literature, there is no consensus in the budget of sediment delivered to the coastal zone varying from 9.3 Gt yr^{-1} to more than 58 Gt yr^{-1} (Milliman and Syvitski, 1992). Estimating the world sediment budget, it is still a challenge either due to lack of data or studies in this field (Vörösmarty et al., 2003). Adding to that, there is a highly variability in sediment supply due to the human and climate change impact.

Changes in the landscape as deforestation, land cover and hydraulic mining are the main factors of increasing sediment supply. On the contrary, channel diversion and dam construction trap sediment decreasing the supply. Estimations show that dammed rivers have 50% of trapping efficiency and the numbers of interventions to control river flow have been increasing (Vörösmarty et al., 2003; Wright and Schoellhamer, 2004).

Sediment supply reduction results in greater sediment transport capacity than supply, this imbalance leads to depletion of sediment pool downstream and thus incision in the channels and later clearing of the estuary (Schoellhamer, 2011; Curtis et al. 2010).

1.1.3 Sediment and ecology

The sediment budget and SSC defines the estuary habitat. Salt marshes are mud flats colonized by macrophytes vegetation located in the intertidal zone of an estuary. They are important to the maintenance of a healthy environment as nursery habitat, coastal stabilization, runoff filtration, trophic support and trapping of sediment (Whitcraft and Levin, 2007). Despite their ecological importance, over time, more than 90% of the original marshes in the San Francisco Bay area were leveed and removed from the intertidal area, in order to give space to agriculture and urban land.

In the last decade, the importance of keeping and restoring these habitats has been discussed again. Several projects of marshes/wetlands restoration are being planned and implemented. However, there is a lack of knowledge about how and if these marshes are going to be able to develop and cope with sea level rise. Kirwan et al. (2010) show that marshes can keep up with conservative projections for sea level rise once the sediment concentration is above 20 mg L^{-1}. The results show that the amount of sediment available determines marshes survival. In an environment with a high concentration of sediment in suspension (30-100 mg L^{-1}), such as in the Sacramento-San Joaquin Delta, the marshes are able to survive and cope with a sea level rise up to 10 mm yr^{-1} (Reed, 2002). This adaptability can be maintained since the relative sea level rise does not exceed 1.2 cm yr^{-1} (Morris, 2002).

SSC also defines habitat for phytoplankton, fishes, and small invertebrates. SSC attenuates light penetration in the water, for the phytoplankton the fine sediment carries nutrients but at the same time decrease the photic zone. In the fishes and small invertebrates, it increases survival chance by hiding from predators.

1.1.4 Spatial-temporal scale

Estuarine systems, as San Francisco Bay-Delta (Fig 1-1), present several processes in various spatial/time scales. This means that the overall behavior of system responds to processes on the order of centimeter that happen in seconds at the same time as processes order of kilometers that take decades or even centuries to conclude. This is called the spatial/temporal scale (De Vriend, 1991). Each spatial scale is linked to a certain temporal scale, and they are grouped in levels. There is a hierarchical arrangement of levels, in which the lower levels contain the high-frequency processes and the higher levels, the lower frequency processes.

The system is forced mainly by tides, river discharge, and the wind and we are interested in variability in the order of days, months, years and multi-year. We focus on 3 main processes timescales to study SSC (turbidity) patterns and sediment budget for tide- (Alviso Slough, South Bay) and event-driven estuary (Sacramento-San Joaquin Delta):

- Days to weeks - tidal asymmetry, peak river discharge.
- Months - reverine seasonal cycle. The rivers that discharge in the Delta present a strong seasonal cycle. Each water year presents a cycle, on average, of 3 wet months and 9 dry months.
- Intra-annual -the dry and wet cycle varies in duration and magnitude from year to year. In addition, the estuary experiences sea level rise and civil works.

1.1.5 Modeling framework

Numerical models are useful tools to investigate and predict SSC patterns and calculate sediment budget since in situ measurements are costly, temporal/spatial limited and cannot forecast. In order to jointly represent the several time and spatial scales aforementioned, it is necessary to model the estuary as a unique system from sediment source to sediment sink. This means, consider in the same domain rivers, estuary, and coastal system. Syvitski et al. (2010) argue that inter-scale communication (such as coupling marshes, estuaries, and rivers) is one of the biggest gaps in the modeling knowledge.

One path to start filling this knowledge gap would be to couple the entire system in a single domain model. This coupling allows propagation of a single forcing variation evenly in the whole domain. It also means that a boundary or input change in the model needs to be done just once, instead of several times for smaller models, saving time and money. On the other hand, it also implies a fairly complex model with many simultaneous processes.

This inability arises from the inherent processes scale, where the spatial and time scales are directly related. So a small scale process (cm-m) will take place in the order of seconds or hours while a large scale (tens km) could take place in the order of decades or even centuries (De Vriend, 1991). This means that it is necessary a model robust enough to guarantee stability for the small and large -scale process and feasible computational time for the large scale when running in a small-scale time step. The models developed so far can be classified into two groups: data-based and process-based models.

Data-based, or behavior-oriented, models are widely applied to predict response to long-term forcing (e.g. sea level rise) (Stive et al., 1998; Dennis et al., 2000; Niedoroda et al., 1995; Karunarathna et al., 2008). To develop this type of model, a large amount of data is needed in order to extract the system behavior from the data. The behavior-oriented approach takes into consideration equilibrium relationships like the equilibrium beach profile (Brunn et al., 1962), and the equilibrium relationship assumed between the channel volume and the tidal prism. Within this classification, several approaches are considered like hybrid models and statistical analysis.

The hybrid models consider dynamics that are designed to predict qualitative behavior by including only predominant processes. They are based on elementary physics but include most empirical knowledge of equilibrium states. Some simplifications are made to define the estuary as a number of morphological boxes which leads to an equilibrium state (Capobianco et al., 1999). One example of this type of model is ASMITA (Aggregated Scale Morphological Interaction between Tidal basin and Adjacent coast) (Wang, 1998).ASMITA divides the estuaries into elements that are described by a single variable volume (box model). This type of model requires a small number of elements to produce useful solutions.

The statistical model is based on statistical analysis of the data available. It is possible to define the oscillation mode of a system applying statistical tools such as bulk statistics (mean, standard deviation, correlation, etc.), empirical orthogonal functions (EOF), canonical correlation analysis (CCA), complex EOF or complex principal component analysis (CPCA), and singular spectrum analysis (SSA), described in Larson et al. (2003), Southgate et al. (2003) and Kroon et al. (2008). The relationships derived from this method are mostly case specific.

Process-based models describe the physical process (waves, currents, sediment transport) through mathematical formulations based on first physical principles such as conservation of mass, momentum, energy (Latteux, 1995; De Vriend and Ribberink, 1996; Lesser et al., 2004). These models do not preserve an equilibrium state but rather describe the physics that may lead to it. This type of model can deal with complex geometry and include several processes as well as reproduce results for idealized models. Recent studies (Dastgheib et al., 2008; Marciano et al., 2005; van der Wegen and Roelvink, 2008) show that applying bed level updating techniques described by (Roelvink, 2006) it is possible to predict long-term evolution using process-based models. Dissanayake et al. (2011) compare the results of a process-based model (Delft3D) with a semi-empirical model (ASMITA) for a long-term morphological change of a tidal inlet. They conclude that in the long term (decades) both models converge to a similar state, however during the adaptation time they present different behaviors.

There are some weak points that should be carefully managed in the process-based model. The first weak point is untested theories, which do not give a correct interpretation of phenomenon, propagating errors (Syvitski, 2007). The second weak point is determining the important processes; since it is computationally impossible to model all the processes of an estuary the choice of which processes to include is extremely important (Werner, 2003). However, making the right choice of processes to include it is possible to have a detailed representation of the transient solution as well as the steady state.

In the scope of the process-based models, the numerical grid where the physical equations will be calculated can be rectangular, curvilinear or unstructured. Even though the rectangular and curvilinear grids are more computationally efficient, they do not fit in all geometries. This is the case for many estuaries, where a complex channel network merges with bays or seas. In the San Francisco Bay case, the unstructured mesh model solves geometry issue of merging the Delta grid to the Bay region.

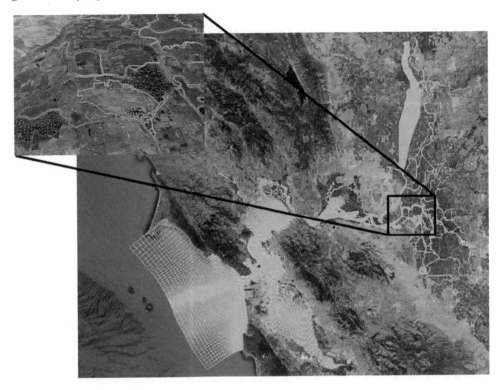

Figure 1-1: Model grid, from the coastal region up to Freeport, in Sacramento River and Vernalis in San Joaquin River. In detail a zoom from Frank's Tract to show the grid flexibility.

There are three widely known unstructured mesh models: TELEMAC-MASCARET (Hervouet, 2007), UnTRIM (Casulli and Walters, 2000) and Delft3D FM (Kernkamp et al., 2010) (Fig 1-1). The reasons for choosing Delft3D FM for this study is based on the CASCaDE II project interests of having the hydrodynamic and morphodynamic model coupled with a water quality/ecology model, Delft-WAQ. This model allows gridding the whole system of rivers, delta, bay and coastal area in a unique grid.

The Delft3D FM is a process-based unstructured grid model and is currently being developed by Deltares. It is a package for hydro- and morphodynamic simulation based on finite volume approach and applies a Gaussian solver. Delft3D FM accounts for 1D, 2D, and 3D schemes. The

2D scheme solves the depth-integrated shallow water equation (hydrostatic). However, when vertical processes are important, there is a 3D scheme, which solves the full Navier–Stokes (or non-hydrostatic) equations. Having this sort of separation, there is a big gain in computational time when only 2D approach is necessary. Delft3D FM generates output for off-line coupling with water quality model Deft3D-WAQ. Delft3D-WAQ is a 3D water quality model framework. It solves the advection-diffusion-reaction equation on a predefined computational grid and for a wide range of model substances. Delft3D-WAQ allows great flexibility in the substances to be modeled, as well as in the processes to be considered.

1.2 Motivation: the CASCaDE II project and BDCP

1.2.1 CASCaDE II project

The Ph.D. project is embedded in the CASCaDE II (Computational Assessments of Scenarios of Change for Delta Ecosystem) project framework (Fig 1-2). The CASCaDE II project is an interdisciplinary and inter-institution project with the objective of better understanding the whole system of San Francisco Bay, Sacramento-San Joaquin Delta, tributaries, rivers and Watershed (BDRW). The following studies and modeling of the areas were done: climate downscaling; watershed water supply; sediment supply; hydrodynamics; phytoplankton dynamics; turbidity and sediment budget (this Ph.D. work); marsh sustainability; prediction of sediment supply; contaminant biodynamic; invasive bivalves and fish ecology.

There are 20 researchers working on the project from American and European institutions such as: United States Geological Survey (USGS), University of California Davis (UC Davis), National Oceanic and Atmospheric Administration (NOAA), Interagency Ecological Program Leading Scientist, UNESCO-IHE, and Deltares.

The project aims are to define and quantify "how will future changes in physical configuration and climate affect water quality, ecosystem process, and key species in the Delta?" The question arises from the new state law that mandates that the Delta ecosystem and a reliable water supply have equal value and are a priority for the water management in California. The Delta is one of the most important water suppliers from California.

The main output from this Ph.D. for the entire project is the calibrated model describing yearly sediment budgets for the benthic group and forecast of turbidity levels for the phytoplankton, contaminants, bivalves, fish, and marshes. So most of the groups are dependents of these results what show the high relevance of the work.

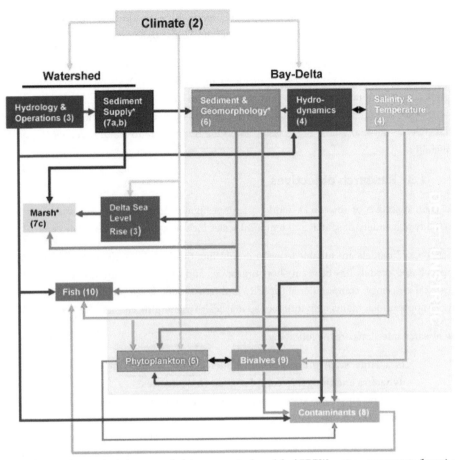

Figure 1-2: Flow chart is depicting an underlying conceptual model of BDRW system components (boxes) and interconnections/informational dependencies (arrows), as we propose to represent them in CASCaDE II. * represents new initiatives in CASCaDE II; all other components were developed or initiated in CASCaDE I and will be enhanced and extended in CASCaDE II. (Marsh accretion and watershed sediment supply are completely new. Sediment models for northern SF Bay were developed in CASCaDE I, but the Delta sediment/geomorphology model is new. Hydrodynamics, salinity, temperature, phytoplankton and bivalve modeling were performed/initiated in CASCaDE I, but the UNSTRUC framework for these is new in CASCaDE II).

1.2.2 BDCP

This research has a high political impact since it developed a tool and assessment of possible impacts to the BDCP (Bay Delta Conservation Plan, www.baydeltaconservationplan.com). The Sacramento-San Joaquin Delta is setting for many water conflicts including industries, agriculture, water supply for population and habitat of endemic species. Increasing pressure of influential sectors demanding freshwater security led the government to rethink previous projects of water diversions resulting in BDCP.

Discussion about a diversion structure construction from Sacramento to South Delta dates back to the 1970s. The last proposal was the Bay Delta Conservation Plan (BDCP). The plan calls for construction of a tunnel; that would link the Sacramento River to Clifton Court pumping station, with 3 intake stations that together have a capacity of 252m³s⁻¹ (9000 cfs).

BDPC will drastically change the Delta flow as the Delta is an important habitat a number of safeguards and restoration projects are included in the BDCP. BDCP consider monitoring fish migration and habitat, restoration and protection of marshes, floodplain, channel margin, riparian habitat, grassland, and wetlands. Most of these systems need sediment to develop, but so far they do not know how big is going to be the impact due to water diversion in sediment availability.

1.3 Research objectives

The main objective of this Ph.D. work is to investigate turbidity level and sediment budget variability at a variety of spatial and temporal scales including the way these scales interact.

Process-based models are suitable to assist in fulfilling this task. Until recently, the application of process-based models has been on short timescales and limited domain, due to the increase in numerical efficiency, computational capacity and new model developments, this type of models may be applied to domains with different spatial scales as merging coastal area, bay, and rivers.

The research questions are the following:

> **Is a large scale process-based model a suitable tool for reproducing sediment dynamics and budget in complex geometry estuaries?**
> **How much in situ data is necessary to develop a calibrated sediment model for complex geometry estuaries?**
> **What do we learn from comparing time/spatial scales of an event-driven system and a tidal-driven system?**
> **To what extent can we predict future scenarios? What is the model applicability?**

We use the San Francisco Bay-Delta as a study case. San Francisco Bay-Delta is a suitable study case because (a) it has been relatively well measured over a long time span. The available data comprises biotic and abiotic data such as water level, current velocities, river discharge, sediment concentration, all sorts of water quality data, marsh evolution and phytoplankton, clams and fish. This amount of data provides a good study case for model validation. (b) It is a proper test case for the model to be implemented since it is an environment with complex geometry, the sediment is a mixture of sand and mud, and is subject to extensive human interference. (c) Apart from anthropogenic relevance, the delta (considered here as the river channel network that is formed inland, before the bay) and surrounding wetlands provide habitat, nursery and trophic support for several endemic species.

The tool aims to breach a gap between abiotic and ecological modeling by providing input to ecological modelers

This research has scientific and management impact. The scientific branch includes assistance in the development of Delft3D FM. The Bay-Delta model is one of the first real case applications of the new numerical model. Another scientific aspect is the detailed analysis of sediment flux and turbidity patterns temporal and spatial scales that are further coupled with ecological models (not the scope of this research). Management wise, this research provided a calibrated model to understand the system, to assist in the Delta operations, surveying campaigns, and forecast impacts. This environment encompasses different and conflicting water uses. As such, the case study provides an excellent opportunity to assess modeling opportunities for estuaries worldwide that experience a similar pressure.

1.4 Outline of the thesis

This thesis is structured in six chapters. In addition to the introduction and conclusion, this thesis contains four other chapters that follow the research questions presented above.

Chapter 2 describes the main system temporal and spatial scales in terms of turbidity, sediment flux and budget in the Sacramento-San Joaquin Delta. The turbidity patterns, spatial distribution, are time dependent. So, it describes the time variation of the spatial turbidity patterns as a step to a chain of models.

Chapter 3 assess the importance of the Delta channel network in the sediment budget and deposition patterns and the relevance of the peak event in the dynamics.

Chapter 4 presents the sediment dynamics applying the same methodology from the previous chapter for a tidal driven estuary, the Alviso Slough. Alviso Slough has different temporal scales and includes the problematic of mercury-contaminated sediment.

Chapter 5 discusses the relative impact future scenarios of sea level rise, management operations by the BDCP and levee failure in the Delta sediment dynamics. The scenario analysis gives the system resilience due to different impacts.

A 2D PROCESS-BASED MODEL FOR SUSPENDED SEDIMENT DYNAMICS: A FIRST STEP TOWARDS ECOLOGICAL MODELING

In estuaries suspended sediment concentration (SSC) is one of the most important contributors to turbidity, which influences habitat conditions and ecological functions of the system. Sediment dynamics differ depending on sediment supply and hydrodynamic forcing conditions that vary over space and over time. A robust sediment transport model is a first step in developing a chain of models enabling simulations of contaminants, phytoplankton and habitat conditions. This work aims to determine turbidity levels in the complex-geometry Delta of San Francisco Estuary using a process-based approach (Delft3D Flexible Mesh software). Our approach includes a detailed calibration against measured SSC levels, a sensitivity analysis on model parameters, and the determination of a yearly sediment budget as well as an assessment of model results in terms of turbidity levels for a single year, Water Year 2011. Model results show that our process-based approach is a valuable tool in assessing sediment dynamics and their related ecological parameters over a range of spatial and temporal scales. The model may act as the base model for a chain of ecological models assessing the impact of climate change and management scenarios. Here we present a modeling approach that, with limited data, produces reliable predictions and can be useful for estuaries without a large amount of processes data.

This chapter is based on:

Achete, F. M.; van der Wegen, M.; Roelvink, D., Jaffe, B.:   A 2D Process-Based Model for Suspended Sediment Dynamics: a first Step towards Ecological Modeling, Hydrol. Earth Syst. Sci., 19, 2837–2857, 2015 doi:10.5194/hess-19-2837-2015

2.1 Introduction

Rivers transport water and sediments to estuaries and oceans. Sediment dynamics will differ depending on sediment supply and hydrodynamic forcing conditions, both of which vary over space and time. The human impact on sediment production dates from 3000 years ago, and has been accelerating over the past 1000 years due to considerable engineering works (Milliman and Syvitski, 1992; Syvitski and Kettner, 2011) estimate that the budget of sediment delivered to the coastal zone varies between 9.3 and 58 Gt per year. Estimating the world sediment budget is still a challenge because of the lack of data and detailed modeling studies (Vörösmarty et al., 2003). In addition, there is considerable uncertainty in hydraulic forcing conditions and sediment supply dynamics due to variable adaptation timescales over seasons and years (such as varying precipitation and river flow), decades (such as engineering works) and centuries to millennia (sea level rise and climate change).

Examples of anthropogenic changes influencing sediment dynamics in river basins and estuaries are manifold, e.g., the San Francisco Bay-Delta (Schoellhamer, 2011) and Yangtze estuaries (Yahg, 1998), and the Mekong Delta (Manh et al., 2014). These three systems are similar in how anthropogenic changes altered sediment supply. After an increase in sediment supply (due to hydraulic mining and deforestation respectively) each had a steep drop in sediment discharge (30% or more) due to reservoir building and further estuarine clearing after depletion of available sediment in the bed. This implies a) continuous change in sediment dynamics and hence sediment budget in the estuary and b) change in sediment availability leading to change in turbidity levels.

Turbidity is a measurement of light attenuation in water and is a key ecological parameter. Fine sediment is the main contributor to turbidity. Therefore suspended sediment concentration (SSC) can be translated into turbidity applying empirical formulations. Besides SSC, algae, plankton, microbes and other substances may also contribute to turbidity levels (ASTM International, 2002). High turbidity levels limit photosynthesis activity by phytoplankton and microalgae, therefore decreasing associated primary production (Cole et al., 1986). Turbidity levels also define habitat conditions for endemic species (Davidson-Arnott et al., 2002). For example, in the San Francisco Bay-Delta estuary, the Delta Smelt seeks regions where the turbidity is between 12-18 NTU to hide from predators (Baskerville and Lindberg, 2004; Brown et al., 2013). Examples of other ecological impacts related to SSC are vegetation stabilization (Morris et al., 2002; Whitcraft and Levin, 2007), and salt marsh survival under sea level rise scenarios (Kirwan et al., 2010; Reed, 2002).

To assess the aforementioned issues, the goal of this work is to provide a detailed analysis of sediment dynamics including a) SSC levels in the Sacramento-San Joaquin Delta (Delta), b) sediment budget and c) translation of SCC to turbidity levels using a two-dimensional horizontal, averaged in the vertical (2DH), process-based, numerical model. The 2DH model solves the 2D vertically integrated shallow water equations coupled with advective-diffusive transport. This process-based model will be able to quantify high-resolution sediment budgets and SSC, both in time (~ monthly/yearly) and space (~10s-100s of m). We selected the Delta area as a case study since the area has been well monitored so that detailed model validation can take place, it hosts endemic species and allow us to use a 2DH model approach.

The Delta and Bay are covered by a large survey network with freely available data on river stage, discharge, and suspended sediment concentration (SSC) and other parameters from the USGS (nwis.waterdata.usgs.gov), Californian Department of Water Resources (http://cdec.water.ca.gov/) and National Oceanic and Atmospheric Administration (http://tidesandcurrents.noaa.gov/). The continuous SSC measurement stations are periodically calibrated using water collected in situ; that is filtered and weighed in the laboratory. In addition, the Bay-Delta system has high resolution (10m) bathymetry available for all the channels and bays (http://www.D-Flow-baydelta.org/).

Regarding ecological value, starting from the bottom of the food web, the Delta is the most important area for primary production in the San Francisco Estuary. The Delta is one order of magnitude more productive than the rest of the estuary (Jassby et al., 2002; Kimmerer, 2004). It is an area for spawning, breeding and feeding for many endemic species of fishes and invertebrates, including some endangered species like delta smelt (Brown et al., 2013), Chinook salmon, spring run salmon and steelhead. Additionally, several projects for marsh restoration in the Delta are planned, and the success of these projects depends on sediment availability (Brown, 2003).

SSC spatial distribution and temporal variability are important information for the ecology of estuaries. However, observations including both high spatial and temporal resolution of SSC are difficult to make, so we revert to using a coupled hydrodynamic-sediment transport models to make predictions at any place and time.

For the first time, a detailed, process-based model is developed for the San Francisco Bay-Delta, to focus on the complex Delta sediment dynamics. From this model, it is possible to describe the spatial sediment (turbidity) distribution and deposition patterns that are important indicators to assess habitat conditions. Seasonal and yearly variations in sediment dynamics and turbidity levels can be used as indicators for ecological modeling (Janauer, 2000). This work fills the gap between the physical aspects (hydrodynamic and sediment modeling) and ecology modeling. Previous work focused on understanding the San Francisco Bay-Delta system through data analysis (Barnard et al., 2013; Manning and Schoellhamer, 2013; McKee et al., 2006; McKee et al., 2013; Morgan-King and Schoellhamer, 2013; Schoellhamer, 2011; Schoellhamer, 2002; Wright and Schoellhamer, 2004, 2005b), while similar work in other estuaries around the world does not give the direct link to ecology (Manh et al., 2014).

2.2 Study area and model

San Francisco Estuary is the largest estuary on the U.S. West Coast. The estuary comprises San Francisco Bay and the inland Sacramento-San Joaquin Delta (Bay-Delta system), which together cover a total area of 2900 km^2 with a mean water depth of 4.6 m (Jassby et al., 1993). The system has a complex geometry consisting of interconnected sub-embayments, channels, rivers, intertidal flats, and marshes (Fig 2-1). The Sacramento-San Joaquin Delta (Delta) is a collection of natural and man-made channel networks and leveed islands, where the Sacramento River and the San-Joaquin River are the main tributaries followed by Mokelumne River (Delta Atlas, 1995). San Francisco Bay has 4 sub-embayments. The most landward is Suisun Bay followed by San Pablo Bay, Central Bay (connecting with the sea through Golden Gate) and, further southward, South Bay.

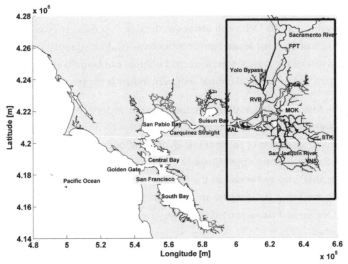

Fig 2-1: Location of the San Francisco Bay-Delta. The black rectangle highlights the Delta, and the red squares indicate measurement stations

Tides propagate from the Golden Gate into the Bay and most of the Delta up to Sacramento (FPT) and Vernalis (VNS) when river discharge is low. Suisun Bay experiences mixed diurnal and semidiurnal tide that ranges from about 0.6 m during the weakest neap tides to 1.8 m during the strongest spring tides. During high river discharge, the 2 psu isohaline is located in San Pablo Bay while, during low river discharge, it can be landwards of Chipps Island (westernmost reach of the black rectangle, Fig 2-1). The topography greatly influences the wind climate in the Bay-Delta system. Wind velocities are strongest during spring and summer with afternoon North-Westerly gusts of about 9 m s^{-1} (Hayes et al., 1984).

San Francisco Estuary collects 40% of the total Californian freshwater discharge. It has a Mediterranean climate, with 70% of rainfall concentrated between October and April (winter) decreasing until the driest month, September (summer) (Conomos et al., 1985). The orographic lift of the Pacific moist air linked to the winter storms and the snow melts in early spring govern this wet (winter) and dry (summer) season variability. This system leads to a local hydrological 'Water Year' (WY) definition from 1st October to 30th September, including a full wet season in one WY.

The Sacramento and San Joaquin Rivers, together, account for 90% of the total freshwater discharge to the estuary (Kimmerer, 2004). The daily inflow to the Delta follows the rain and snowmelt seasonality, with average dry summer discharges of 50-150 m^{3}s^{-1} and wet spring/winter peak discharges of 800-2500 m^3s^{-1}. The seasonality and geographic distribution of flows lead to several water issues related to agricultural use, habitat maintenance, and water export. On a yearly average 300 m^3s^{-1} of water is pumped from South Delta to southern California. The pumping rate is designed to keep the 2 psu (salinity) line landwards of Chipps Island avoiding salinity intrusion in the Delta, allowing a 2DH modeling approach.

The hydrological cycle in the Bay-Delta determines the sediment input to the system, thus biota behavior. McKee (2006) and Ganju and Schoellhamer (2006) observed that a large volume of sediment passes through the Delta and arrives at the Bay in pulses. They estimated that in 1 day approximately 10% of the total annual sediment volume could be delivered and in extremely wet years and up to 40% of the annual total sediment volume can be delivered in 7 days. During wet months, more than 90% of the total annual sediment inflow is supplied to the Delta.

The recent Delta history is dominated by anthropogenic impacts. In the 1850`s hydraulic mining started after placer mining in rivers became unproductive. Hydraulic mining remobilized a huge amount of sediment upstream of Sacramento. By the end of the nineteenth century, the hydraulic mining was outlawed leaving approximately $1.1x10^9$ m^3 of remobilized sediment, which filled mud flats and marshes up to 1 meter in the Delta and Bay (Wright and Schoellhamer, 2004; Jaffe et al., 2007). At the same time of the mining prohibition, civil works such as dredging and construction of levees and dams started, reducing the sediment supply to the Delta (Delta Atlas, 1995; Whipple et al., 2012).

Typical SSC in the Delta ranges from 10 to 50 mg L^{-1}, except during high river discharge when SSC can exceed 200 mg L^{-1} reaching values over 1000 mg L^{-1} (McKee et al., 2006; Wright and Schoellhamer, 2005). Sediment budget reflects the balance between storage, inflow and outflow of sediment in a system. Studies based on sediment inflow and outflow estimated that about two-third of the sediment entering the system deposits in the Delta (Schoellhamer et al., 2012; Wright and Schoellhamer, 2005). The remaining third is exported to the Bay, and represents on average 50% of the total Bay sediment supply (McKee et al., 2006); the other half comes from smaller watersheds around the Bay (McKee et al., 2013).

Several studies have been carried out to determine sediment pathways and to estimate sediment budgets in the Delta area (Gilbert, 1917; Jaffe et al., 2007; McKee et al., 2006; McKee et al., 2013; Schoellhamer et al., 2012; Wright and Schoellhamer, 2005). These studies were based on data analysis and conceptual hindcast models. Although the region has a unique network of surveying stations, there are many channels without measuring stations. This might lead to incomplete system understanding and knowledge deficits for the development of water and ecosystem management plans. The monitoring stations are located in discrete points hampering spatial analysis. Also, the impact of future scenarios related to climate change (i.e. sea level rise and changing hydrographs) or different pumping strategies remains uncertain.

2.2.1 Model description

Structured grid models such as Delft3D and ROMS (Regional Oceanic Modeling System) have been widely used and accepted in estuarine hydrodynamics and morphodynamics modeling including studies of the San Francisco Estuary (Ganju and Schoellhamer, 2009; van der Wegen et al., 2011). In all of these studies, the Delta was schematized as 2 long channels because the grid is not flexible, which would have allowed efficient 2D modeling of the rivers, channels and flooded island of the system together with the Bay.

In cases with complex geometry unstructured grids or a finite volume model is more suitable. There are three widely known unstructured grid models: (1) the TELEMAC-MASCARET (Hervouet, 2007), (2) the UnTRIM (Casulli and Walters, 2000; Bever and MacWilliams, 2013) and (3) Delft3D FM (Kernkamp et al., 2010). The two first models are purely triangle based and are not directly coupled (yet) with sediment transport and/or water quality and ecology models.

The numerical model applied in this work is Delft3D Flexible Mesh (Delft3D FM). Delft3D FM allows straightforward coupling of its hydrodynamic modules with a water quality model, Delft-WAQ (DELWAQ), which gives flexibility to couple with a habitat (ecological) model. Delft3D FM is a process-based unstructured grid model developed by Deltares (Deltares, 2014). It is a package for hydro- and morphodynamic simulation based on a finite volume approach solving shallow-water equations applying a Gaussian solver. The grid can be defined in terms of triangles, (curvilinear) quadrilaterals, pentagons, and hexagons, or any combination of these shapes (Fig 2-2). Orthogonal quadrilaterals are the most computationally efficient cells and are used whenever the geometry allows. Kernkamp (2010) and the Delft3D FM manual (Deltares, 2014) describe in detail the grid aspects and numerical solvers.

The Bay area and river channels are defined by consecutive curvilinear grids (quadrilateral) of different resolution. Rivers discharging in the Bay, and channel junctions are connected by triangles (Fig 2-2). The average cell size ranges from 1200m x 1200m in the coastal area, to 450x600m in the Bay area, down to 25x25m in Delta channels. In the Delta, each channel is represented by at least 3 cells in the across-channel direction (Fig 2-2). The grid flexibility allows including the entire Bay-Delta in a single grid containing 63.844 cells of which about 80% are rectangles which keep the computer run times at an acceptable level. It takes 6 real days to run 1 year of hydrodynamics simulation and 12 hours to run the sediment module on an 8-core desktop computer. Besides the triangular grid orthogonality issues, using an entirely triangular grid for a 1-year simulation would increase run times from ~72 to ~192 hours.

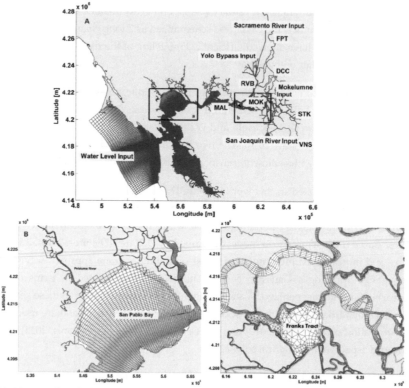

Figure 2-2: Numerical mesh for the Delft3D FM model. Red dots indicate the calibration stations. (http://san-francisco-bay-delta-model.unesco-ihe.org/). Zoom ins of the computational grid, A) San Pablo Bay connecting to the Petaluma and Napa Rivers, B) Delta channels and Franks Tract.

We assume that the main flow dynamics in the Delta is 2D, which doesn't account for vertical stratification. The Delta does not experience salt-fresh water interactions due to the pumping operations, and we assume that temperature differences between the top and bottom of the water column do not govern flow characteristics. Delft3D FM generates hydrodynamic output for off-line coupling with water quality model DELWAQ (Deltares, 2004). Off-line coupling enables faster calibration and sensitivity analysis. Delft3D FM generates time series of the following variables: cell link area; boundary definition; water flow through cell link; pointers that give information about neighbors' cells; cell surface area; cell volume; and shear stress file, which is parameterized in Delft3D FM using Manning's coefficient. Given a network of water levels and flow velocities (varying over time) DELWAQ can solve the advection-diffusion-reaction equation for a wide range of substances including fine sediment, the focus of this study. DELWAQ solves sediment source and sink terms by applying the Krone-Parteniades formulation for cohesive sediment transport (Ariathurai and Arulanandan, 1978; Krone, 1962) (Eq.2-1, Eq.2-2).

$$D = w_s * c * (1 - \tau_b/\tau_d), \text{ which is approximated as } D = w_s * c \qquad (2\text{-}1)$$

$$E = M * (\tau_b/\tau_e - 1) \qquad for\ \tau_b > \tau_e \qquad (2\text{-}2)$$

where D is the deposition flux of suspended matter (mg m^{-2}s^{-1}), w_s is the settling velocity of suspended matter (m s^{-1}), c is the concentration of suspended matter near the bed (mg m-3), τ_b is bottom shear stress (Pa), and τ_d is the critical shear stress for deposition (Pa), The approximation is made assuming, like Winterwerp et al. (2006), that deposition takes place regardless of the prevailing bed shear stress. τ_d is thus considered much larger than τ_b and the second term in parentheses of Equation 1 is small and can be neglected. E is the erosion rate (mg m^{-2} s^{-1}), M is the first order erosion rate (mg m^{-2} s^{-1}), and τ_e is the critical shear stress for erosion (Pa).

2.2.2 Initial and boundary conditions

The Bay-Delta is a well-measured system; therefore, all the input data to the model are in situ data. Initial bathymetry has 10 m grid resolution, which is based on an earlier grid (Foxgrover et al., http://sfbay.wr.usgs.gov/sediment/delta/), modified to include new data by Wang and Ateljevich (http://baydeltaoffice.water.ca.gov/modeling/deltamodeling/modelingdata/DEM.cfm) and further refined. The bathymetry is based on different data sources including bathymetric soundings and lidar data. The hydrodynamic model includes real wind, which results from the model described by (Ludwig and Sinton, 2000). The wind model spatially interpolates hourly data from more than 30 meteorological stations into regular 1 km grid cells. Levees are included in the model, and temporary barriers are inserted to mimic a typical operating schedule as determined by the California Department of Water Resources (http://baydeltaoffice.water.ca.gov/sdb/tbp/web_pg/tempbsch.cfm).

The hydrodynamic model has been calibrated for the entire Bay-Delta system (see appendix A and http://www.D-Flow-baydelta.org/). Initial SSC was set at 0 mg L^{-1} over the entire domain because the model is initiated during a dry period when SSC is low, and the initial condition rapidly dissipates. The initial bottom sediment is mud at places shallower than 5 meters below Mean Sea Level (MSL) including intertidal mud flats, and sand at places deeper than 5 meters below MSL, which are primarily the channel regions. This implies that the main Delta channels such as the Sacramento, San Joaquin, and Mokelumne are defined as sandy with few mud patches. DELWAQ does not compute morphological changes or bed load transport.

In this study, we applied 5 open boundaries. Water levels at the seaward boundary are based on hourly measurements from the Point Reyes station (tidesandcurrents.noaa.gov/). The other four landward boundaries are river discharge boundaries at the Sacramento River (Freeport), Yolo Bypass (upstream water diversion from Sacramento River), San Joaquin River and Mokelumne River (Cai et al., 2014). Studies show that Sacramento River accounts for 85% of the total sediment inflow to the Delta, while the San Joaquin River accounts for 13% (Wright and Schoellhamer, 2005), so it is reasonable to apply 2 sediment discharge boundaries at Sacramento and San Joaquin River. All river boundaries have unidirectional flow and are landward of tidal influence.

The river water flow hourly input data at the Sacramento River at Freeport (FPT), the San Joaquin River near Vernalis (VNS) and Yolo Bypass (YOLO) were obtained from the California Data Exchange Center website (cdec.water.ca.gov/) (Fig 2-3). The sediment input data, for both input stations FPT and VNS, and calibration stations S Mokelumne R (SMR), N Mokelumne R (NMR), Rio Vista (RVB), Mokelumne (MOK), Little Potato Slough (LPS), Middle River (MDM), Stockton (STK), Mallard Island (MAL) (Fig 2-3), was obtained by personal communication from USGS Sacramento; this data is part of a monitoring program (http://sfbay.wr.usgs.gov).

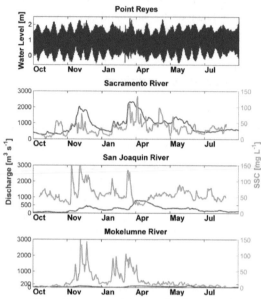

Figure 2-3: Input boundary conditions. The top panel is water level at Point Reyes. The lower 3 panels show discharge in a blue dashed line and SSC in a solid green line for Sacramento River at FPT, San Joaquin River at VNS and Mokelumne River at Woodbridge, respectively.

Since 1998, USGS has continuous measuring stations for sediment concentration which is derived from backscatter sensors (OBS) measurements every 15 minutes and are calibrated approximately monthly with bottle samples (Wright and Schoellhamer, 2005). This type of sensor converts scattered light from the particles to the photocurrent, which is proportional to SSC. To define the rating curve it is necessary to sample water, filter it and weigh the filter. However, in some locations the cloud of points, when correlating photocurrent and filtered weight, shows a large scatter. Large scatter leads to errors in converting photocurrent to SSC. The causes for errors include variation in particle size, particle desegregation (cohesiveness, flocculation, organic-rich estuarine mud); particle shape effects, and sediment concentration effects (Downing, 2006; Gibbs and Wolanski, 1992; Kineke and Sternberg, 1992; Ludwig and Hanes, 1990; Sutherland et al., 2000). Wright and Schoellhamer (2005) showed that for the Sacramento-San Joaquin Delta these errors can sum up to 39% when calculating sediment fluxes through Rio Vista.

In this work, we modeled the 2011 water year - 1st October 2010 to 30th September 2011. First, we ran Delft3D FM for this year to calculate water level, velocities, cell volume and shear stresses. Then, the 1-year hydrodynamic results were imported in DELWAQ, which calculated SSC levels.

The SSC model results are compared to in situ measured SSC data. The calibration process assesses the sensitivity of sediment characteristics such as fall velocity (ws), critical shear stress

(τ_{cr}) and erosion coefficient (M). The model outputs are the spatial and temporal distribution of SSC (turbidity), yearly sediment budget for different Delta regions, and the sediment export to the Bay.

2.3 Results

Our focus is to represent realistic SSC levels capturing the peaks, timing and duration, and to develop a sediment budget to assess sediment trapping in the Sacramento-San Joaquin Delta, (Fig 2-1, highlighted by the black rectangle). Throughout the following sections, the results are analyzed in terms of tide-averaged quantities by filtering data and model results to frequencies lower than 2 days. We applied a Butterworth filter with a cut-off frequency of $1/30h^{-1}$ as presented in Ganju and Schoellhamer (2006).

2.3.1 Calibration

The results shown below are the derived from an extensive calibration process where the different sediment fractions parameters (ws, τ_{cr} and M) were tested. The first attempt applied multiple fraction settings presented in previous works (Ganju and Schoellhamer, 2009a; van der Wegen et al., 2011). However, tests with a single mud fraction proved to be consistent with the data, representative of the sediment budget and allow a simpler model setting and a better understanding of the SSC dynamics. In addition, with a single fraction it was possible to reproduce more than 90% of the sediment budget for the Delta when compared with the sediment budget derived from in situ data.

The best fit of the calibration process (uRMSE=1 and skill=0.8) for the entire domain was obtained in the standard run, which has ws of 0.25 mm s^{-1}, τ_{cr} erosion of 0.25 Pa and M of 10^{-4} kg $m^{-2}s^{-1}$. The initial bed sediment availability is defined by 1 mud (shoals) and 1 sand (channels) fraction. The analysis present below is based in the standard run, and the sensitivity analysis varies the 3 parameters using the standard run as a mid-point.

2.3.2 Suspended Sediment Dynamics (the water year 2011)

The 2011 WY simulation reproduces the SSC seasonal variation in the main Delta regions such as the North (Sacramento River) represented by Rio Vista station (RVB); the South (San Joaquin River) represented by Stockton (STK); Central-East Delta represented by Mokelumne station (MOK) and Delta output represented by Mallard Island (MAL) (Fig 2-4).

All stations clearly reproduce SSC peaks during high river flow from November to July, and lower concentrations during the remainder of the year (apart from MAL during the July-August period). The good representation of the peak timing indicates that the main Delta discharge event is reproduced by the model as well as the periods of Delta clearance. These two periods are critical for ecological models, and a good representation generates robust input to ecological models. The differences found between the model and data are further discussed in appendix B.

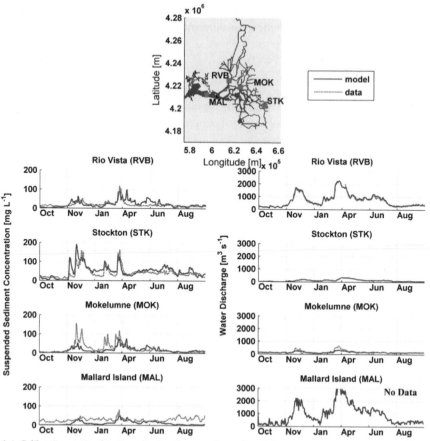

Figure 2-4: Calibration station locations (top) and comparison of model outputs and measured data. Left panels show SSC calibration and right panels the show discharge. Data are dashed red lines, and model results are solid blue lines. Note that in the discharge plots of RVB and STK the data line is behind the model line.

2.3.3 Sensitivity analysis

2.3.3.1 Sediment fraction analysis

We considered one fraction for simplicity and because it reproduces more than 90% of the sediment budget throughout the Delta as well as the seasonal variability of SSC levels. Although more mud fractions considerably increase running time, several tests with multiple fractions were done to explore possibilities for improving the model results.

Including heavier fractions changes the peaks timing and also lowers the SSC curve. Comparing the standard run (ws=0.25 mm s^{-1}, T=0.25 Pa, M=10^{-4} kg m^{-2} s^{-1} and bottom composition with mud available shallower than 5 meters) to another run using 15% of a heavier fraction (ws=1.5 mm s^{-1}) and 30% of a lighter fraction (ws=0.15 mm s^{-1}), showed that the peak magnitudes were underestimated but the first peak timing is closer to the data and the spurious peak mid-May is lower.

To be able to find a single best parameter setting a sensitivity analysis was done varying the main parameters in the Krone-Parteniades formulation (Ariathurai and Arulanandan, 1978; Krone, 1962)(Table 2-1). RVB and MAL are more affected by the tidal wave than STK and regarding sediment flux; these tests show that RVB and MAL are more sensitive to parameter change than STK (Fig 2-5). The model results are most sensitive to the critical shear stress for erosion and least sensitive to the erosion coefficient. Analyzing the time series, one concludes that in stations where the fluxes are higher, the change in critical shear stress is less important, since during most of the time the shear stress is already greater than any given critical shear stress.

Table 2-1: Parameters set of sensitivity analysis.

Parameters		Minimum	Maximum
Standard		w = 0.25; Tau = 0.25; M=1*10⁻⁴	
Fall velocity	ws (mm s⁻¹)	0.15	0.38
Critical shear stress	τ_{cr} (Pa)	0.125	0.5
Erosion Coefficient	M (kg m²s⁻¹)	2.5*10⁻⁵	1*10⁻²

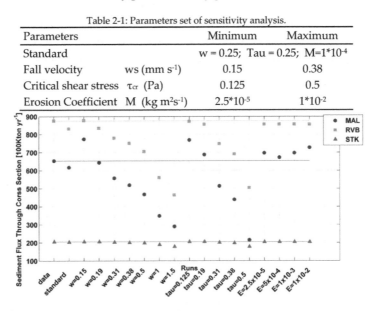

Figure 2-5: Sensitivity analysis for sediment flux at RVB on the Sacramento River (green squares), at STK on the San Joaquin River (red triangles) and at MAL where the Delta meets the Bay (blue circles). The colored lines indicate the data values.

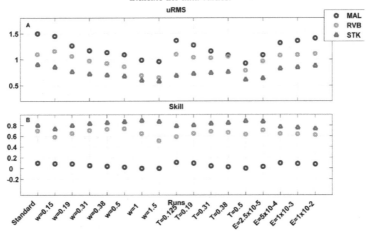

Figure 2-6: Statistical metrics for sensitivity runs. (a) Unbiased Root Mean Square and (b) Skill. On the x-axis are the different runs. Colored symbols are stations RVB (green square), STK (red triangle) and MAL (blue circle).

$$uRMSe = \left(\frac{1}{N}\sum_{i=1}^{N}\left[(X_{mi} - \overline{X_m})(X_{oi} - \overline{X_o})\right]^2\right)^{0.5}$$
(2-3)

Where N is the time series size, X is the variable to be compared, in this case SSC, and $\overline{X}$ is the time-averaged value. Subscript m and O represent modeled and observed values, respectively.

We quantify error using two metrics, the unbiased Root Mean Square Error (uRMSE, Fig 2-6) and Skill (Skill, Fig 2-6) (Bever and MacWilliams, 2013). The uRMSE indicates the variability of the model relative to the data and is 0 when the model and data have equal variability.

Skill is a single quantitative metric for model performance (Willmott, 1981). When skill equals 1, the model perfectly reproduces the data. The 2 metrics were evaluated at RVB, STK, and MAL, representing respectively Sacramento River, San Joaquin River and Delta output.

$$Skill = 1 - \left[\sum_{i=1}^{N}|X_{mi} - \overline{X_{oi}}|^2\right] / \left[\sum_{i-1}^{N}(|X_{mi} - \overline{X_o}| + |X_{oi} - \overline{X_o}|)^2\right]$$
(2-4)

The choice of the standard run analyzed throughout the paper comes from this analysis as well as the budget analysis. We note that both uRMSE and Skill varies up to 50% over the different runs.

2.3.3.2 Initial bottom composition

To study the importance of initial bottom sediment availability we considered 2 cases; one excluding sediment (no sediment available at the bed) and the other with mud at places shallower than 5 meters below MSL, the same setting as the standard run.

We did some tests varying the 5 m threshold. From 3 to 10 meters the final results are all similar. However, allowing mud availability in the channels deeper than 10 meters starts to affect the SSC levels. Time series of SSC comparing the 2 cases show that bottom composition has virtually no influence on SCC after the first couple of days. This result also applies to different mud fractions availability and suggests it may be possible to model accurately less-measured estuaries where virtually no bottom sediment data is available.

Another test shows that it is better to initialize the model with no sediment at bed than with mud available in the entire domain. Initializing the channels with loose mud generates unrealistically high SSC levels through the years, which can take up to 5 years to be reworked.

2.4   Discussion

In the previous section, we presented the model calibration, a normal practice in the modeling process. In this section, we discuss the new insights that were derived from the model results. Although these insights are specific to the San Francisco Bay-Delta system, the same approach can be applied to other estuaries and deltas. The model produces detailed sediment dynamics and the main paths that sediment is transported in the Delta. Sediment flux calculations define the sediment dynamics while gradients in sediment describe the sediment distribution and deposition pattern in the Delta. We also discuss the daily and seasonal variation of turbidity levels.

2.4.1 Spatial sediment distribution

We start the analysis by exploring the general Delta behavior. During dry periods SSC in the entire Delta is low (<20 mg L^{-1}) and the Delta water is relatively clear. The current model results confirm that the Sacramento River is the main sediment supplier into the Delta (Wright and Schoellhamer, 2004; Schoellhamer et al., 2012). Sacramento River peak flow fills the North and partially fills the Central/East Delta with sediment. However, the rest of the Delta has quite low levels (~ 20 mg L^{-1}) of SSC all year long. Passing Vernalis (VNS), San Joaquin River main branch flows to the East. However the SSC peak reaches no much further than STK. The West branch goes toward the water pumping stations where the sediment is pumped out of the system. This behavior results in very low SSC in the South Delta (Old River and Franks Tract) region, which are deposition areas.

Three Mile Slough (TMS) and the Delta Cross Channel (DCC) connect the Sacramento River with the Central and Eastern Delta. Model results show that together they carry 60 Kt per year of sediment southward. DCC operation controls SSC levels in the Eastern/Central Delta to a large extent. To show the importance of the DCC, we run the model twice, once with the DCC always open and once always closed. When the DCC is open, high SSC Sacramento river water (~150 mg L^{-1}) flows towards the Mokelumne River and Eastern Delta increasing the overall SSC in the area. When it is closed, SSC levels in the Central and Eastern delta are about 30 mg L^{-1} lower than in the previous case (Fig 2-7). The effect of opening the DCC can be observed in the SSC level at the San Joaquin River from the MOK station seawards. In the Sacramento River, the opening decreases SSC levels, by about 10 mg L^{-1} and affects the river SSC all the way to Mallard Island (Fig 2-7).

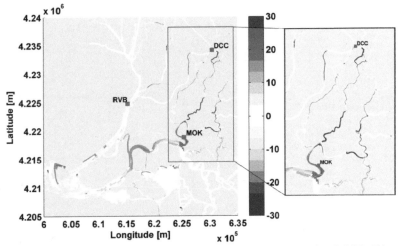

Figure 2-7: Anomaly of a SSC (mg L^{-1}) snapshot between runs with open closed DCC. This pattern is representative in time as well. The right panel is a zoom in between the DCC and MOK (black rectangle). Red shades represent regions where the SSC level is higher in the open than the close scenario, the blue shades where it was lower.

During peak river discharge, Sacramento River sediment reaches Mallard Island in approximately 3 days, Carquinez Straight in 5 days, and the Golden Gate Bridge in approximately 10 days. This timing is proportional to river discharge. However, from Mallard Island seawards this estimate is inexact due to the 2D approximation. San Joaquin River sediment remains largely trapped in the southern Delta. The flooded islands, breached levees like Franks Tract, present a different behavior. During the entire year, the SSC levels are below 15 mg L^{-1} and the peak river discharge signal does not affect them.

Sediment flux is a useful tool for a quantitative and qualitative analysis of the sediment pathways, and its derivative gives sedimentation/erosion patterns. Sediment flux is defined by the product of water velocity (U), times Cross-sectional area (A) times SSC (C) (Eq. 2-5).

$$F_{sed} = U * A * C \qquad\qquad (2\text{-}5)$$

The yearly sediment flux through FPT from model results is 1132 Kt yr^{-1} (thousand metric tons per year) and 1096 Kt yr^{-1} from data. Farther seaward on the Sacramento River at RVB the sediment flux is 832 Kt yr^{-1} (994 Kt yr^{-1}, data). Sediment flux at MAL is 617 Kt yr^{-1} (654 Kt yr^{-1}) (Fig 2-8). We calculate that 30Kt yr^{-1} of Sacramento River sediment flow to the Eastern Delta through the DCC, and 30 Kt yr^{-1} through TMS and 20 Kt yr^{-1} from Georgina Slough. The San Joaquin River carries 490 Kt yr^{-1} (498) through VNS, and at STK 205 Kt yr^{-1} (190 Kt yr^{-1}). An estimated 100 Kt yr^{-1} is exported through pumping. To close the system in Central Delta, the flux through JPT is 126 Kt yr^{-1} (no data) and at the DCH approximately 0 (no data) (Fig 2-8).

Seaward from MAL considerable salt-freshwater stratification takes place in the water column. These 3D effects are not captured by our 2DH approach and model results in this region are inaccurate. Therefore, Fig 2-8 shows preliminary sediment flux to the Bay by a dashed line.

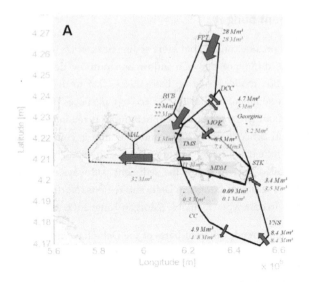

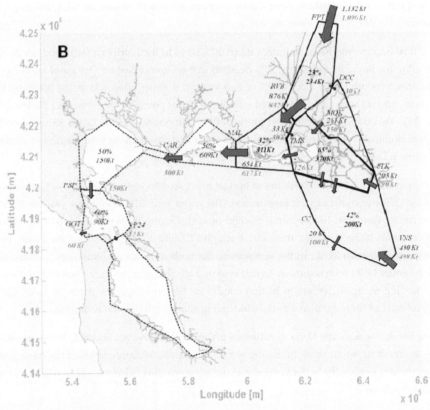

Figure 2-8: Water discharge (A) and sediment flux (B) pathway models. The arrows represent the water (A) and sediment (B) fluxes through the cross sections. Area of the arrow is proportional to the flux. Fluxes from data are in red and from the model are in blue. Inside each polygon are the trapping efficiency and deposition volume for the area. The Bay portion is dashed because the model is 2D and 3D processes occur in that region.

2.4.2 Sediment budget

From the previous section, one can see that more sediment enters (~1600 Kt yr⁻¹) than leaves (~600 Kt yr⁻¹) the Delta. The difference between inflow and outflow deposits in the Delta. Jaffe et al.(2007) developed a box model based on bathymetry data to define sediment budget of the Delta and Bay to define sediment availability for ecology purposes. The model results agree with data estimations that about two third of the sediment input is retained in the Delta (Schoellhamer et al., 2012; Wright and Schoellhamer, 2005), and retention is consistent throughout the years (Cappiella et al., 1999; Jaffe et al., 1998; Wright and Schoellhamer, 2004). Because the Delft3D FM model provides a detailed description of the sediment pathways, it is possible to understand further and describe the sediment budget in Delta sub-regions (North, Central, and South) and to compare model results to data when available (Morgan-King, 2012, personal communication).

Besides the overall spatial trend, different parts of the Delta have different trapping efficiencies. The Northern Delta (the least efficient) traps ~ 23%; Central/Eastern Delta traps 32%, Central/Western 65%, and the most efficient region, the Southern Delta, traps 67% of the sediment input. The highest trapping efficient regions are where islands inundated through levee breaching (Wright and Schoellhamer, 2005).

Of the total Sacramento River sediment input 40% stays in the Northern Delta and about 40% is exported to the Bay. The remaining 20% deposits in the Central/Eastern Delta and only 2% travel all the way to South Delta. About 70% of San Joaquin sediment deposits in the Southern Delta, 10% go to central Delta, 15% is exported via Clifton Court pumping facilities, and 5% is exported to the Bay. This transport is reflected in the bottom composition of the Delta. Sacramento River sediment dominates the Northern and Central Delta and San Joaquin River sediment dominates the Southern Delta bottom composition (Fig 2-9).

It is enlightening to divide the sediment budget analysis into wet and the dry seasons since the Delta has different dynamics for each season. The water year 2011 was a wet year, with the wet season lasting from mid-January until the end of May. During the wet period, 60% of the yearly sediment input budget entered the Delta through FPT and VNS and 70% of the yearly budget was exported through MAL. In the wet season, the high river water discharges and SSC pulses flush the entire Delta with sediment. In this season, high SSC gradients are observed in the plume fronts leading to rapid changes in habitat conditions for many species. After the front the high SSC level can last for more than a month, indicating changing in habitat conditions

During the dry season, the Delta experiences lower river discharges and SSC levels resulting in lower sediment transport rates. In the dry season, SSC levels are more uniform not having peaks. During the dry season, the water is clear and the advective flux is lower, which will be discussed in the next section.

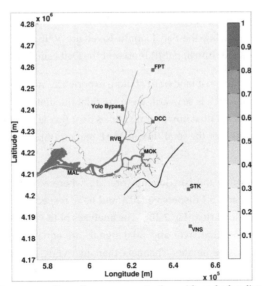

Figure 2-9: Sediment bottom composition after one year, starting with no bed sediment available. Red shades indicate the dominance of Sacramento River sediments and white shades dominance of San Joaquin River sediments. The black line highlights where this separation occurs.

2.4.3 Sediment flux analysis

SSC peaks at FPT can be tracked down the estuary. At the RVB station the SSC peak follows s dynamic as observed at FPT; however, this behavior does not apply for the entire Delta. Schoellhamer and Wright (2005) observed that the river signal is attenuated through the estuary. This attenuation can be understood by analyzing changes in the dominant sediment flux component.

Dyer (1974) decomposed the tidally averaged fluxes in three main components: tidal mean, the advective term; tidal fluctuation, the dispersive term; and the Stokes drift. This decomposition was possible considering that the measured valued is the sum of a tidally mean component [x], and a fluctuating component $x^{\wedge'}$, so $x=[x]+x^{\wedge'}$, substituting in Eq. 2-5 and simplifying the small contribution terms, three main terms remain (Eq.2-6). The first term of Eq. 2-6 is the advective term, the river sediment flux calculated as the product of the mean discharge, area, and concentration; the second term is the dispersive sediment flux that accounts for tidal pumping of sediment. The 2 first terms account for more than 95% of the sediment flux. The remaining sediment flux is from the third term, Stokes drift, which is the transport due to a variation in the cross-sectional area. The other terms are very small representing less than 5% of the total flux, therefore, disregarded for this analysis.

$$[F] = [U][A][C] + [[U'[A]C']] + [[U'A'[C]]]$$ (2-6)

The model allows for a detailed temporal and spatial analysis of the three flux components. The temporal analysis is done for the whole year and for the wet and dry seasons separately. For the spatial analysis, we defined 4 stations for each river where the first station is dominated by the river flux and the last experiences a mix of tidal and river fluxes. The stations follow the

Sacramento River, starting with FPT, followed by RVB down to Mallard Island where the Delta joins the Bay. Stations following the San Joaquin River are VNS, STK, and MOK. Three Mile Slough (TMS) and San Joaquin Junction (SJJ) represent the Delta smaller channels.

Sacramento River at FPT, the most landward station, experiences no tidal influence, so the flux is purely advective. At RVB, which is seaward, there are tidal fluctuations and the dispersive flux is responsible for 22% of the total flux; however no Stokes drift flux is present (Fig 2-10). In contrast, Stokes drift accounts for 33% of the total flux in MAL station implying that tides have a bigger influence in this region.

An analogy can be drawn to the San Joaquin branch, where VNS and STK experience only advective terms. At MOK and SJJ dispersive (20% and 63%, respectively) and Stokes flux (5 and 11%) start to influence the total flux (Fig 2-10). The analyzes of the 3 different flux components in smaller Delta channels show that river and tidal signals are equally important. The river peak signal is less important inside smaller channels than in rivers. At TMS, the dispersive flow accounts for 60% of the total flux.

The flux analyzes show that there is no change in the Delta net circulation when comparing wet and dry seasons. There is not a major change in the flux direction when comparing the seasons. However, there is a change in the importance of each flux component.

Figure 2-10 shows that dispersive flux and Stokes drift relative contributions vary seasonally. When river discharge is high, the relative contribution of dispersive flux and is lower than during low flow conditions. This pattern is more apparent at stations where the river signal is stronger. At RVB, the dispersive flux contribution is about 15% during the wet season and 26% in the dry season. The same applies for MAL and STK. In smaller channels, like TMS and SJJ, the dispersive flux seasonal variation is milder, varying about 10%, from 55% in the wet season and 65% in the dry season. In the dry season the change in flux contributions, from advective to dispersive and Stokes drift, leads to a lower net export of sediment from the Delta, even though the concentrations in the Delta are only about 30mg L^{-1}.

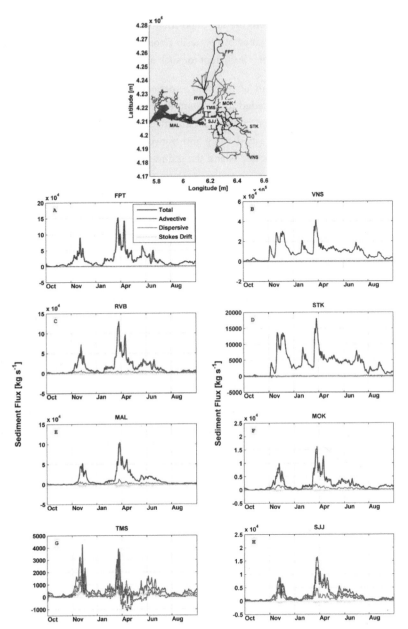

Figure 2-10: Sediment flux calculations for several stations within the Delta. Figs A, C, E, and G show the sediment flux change following the Sacramento branch and B, D, F and H following the San Joaquin branch. The total flux is represented in magenta, advective flux in blue, dispersive flux in red, and Stokes drift in green. The total and advective sediment fluxes are the same at FPT and VNS. Positive is seaward.

2.4.4 Sediment deposition pattern

The flux changes from completely advective to dispersive and Stokes drift sheds some light on the Delta deposition areas. The places where the dispersive flux starts to play a role, near RVB and MOK, are the same places where the net deposition is observed (Fig 2-11). Other locations where considerable sedimentation takes place are in flooded islands areas, such as Frank Tract and the Clifton Court. The 2D model is sufficient for such areas (Fig 2-11).

The San Joaquin River downstream of Stockton experiences high deposition. This finding is confirmed by constant dredging needed to maintain the Stockton navigation channel. The river discharge modulates the deposition pattern in the main channels. In the Sacramento River deposited sediment is gradually washed away and transported to the mud flats at the channel margins, until the next peak. At flooded island the sedimentation process is gradual and steady, erosion is not observed in these areas.

Deposition is primarily observed during the wet and dry season. Some exceptions occur in small bends in the Sacramento River that are erosional during the wet season and depositional during the dry season. The deposition pattern provides insight into the best areas for marsh restoration.

2.4.5 Turbidity

So far the discussion presented is in terms of SSC levels for the standard run, budgets and fluxes, while ecological analysis is often based on turbidity levels. SSC and turbidity are correlated by rating curves as log10 (SSC) = a*log10 (Turb) +b, where a and b are local parameters empirically defined for each Delta area. For the Northern area a=0.85 and b=0.35; Central/Western area a=0.91 and b=0.29, Central/Eastern a=0.72 and b=0.26; Southern a=1.16 and b=0.27; Eastern a=0.914 and b=0.29 (USGS Sacramento, personal communication 2014).

In this section, we present average values for turbidity within a specific Delta region as well as its seasonal and daily variations (Fig 2-12). Generally, the mean turbidity levels and spatial variations are higher during the wet season than during the dry season. During the wet season, the Southern area had the highest mean value (50 NTU), and deviation (15 NTU), caused by a combination of large sediment supply and low flow velocities. The Northern region is the second most turbid area (45±10 NTU), where sediment transported by the Sacramento River flows in the channels, increasing the turbidity levels. The Central West region is the least turbid area (5±2 NTU) and, as previously shown, it has the highest trapping efficiency of the entire Delta. In the dry season, the mean turbidity daily variation decreases in the entire Delta. The opening of the DCC during the dry season lets sediment from the Sacramento River enter these areas, increasing the mean turbidity level. The spatial distribution of the most turbid areas is the same as in the wet season. The daily deviation is mostly proportional to the turbidity level and to the distance from the sea. In the Southern and Western areas the daily variation is higher during the dry season. It shows that there is a strong tidal signal in these parts of the Delta.

The DCC and GLS channels that connect the Sacramento and San Joaquin Rivers are important bridges to export sediment from the Sacramento River to Eastern Delta. The smaller channels of the network play a minor role in the Delta sediment budget because the discharges in these channels are considerably smaller than in the rivers.

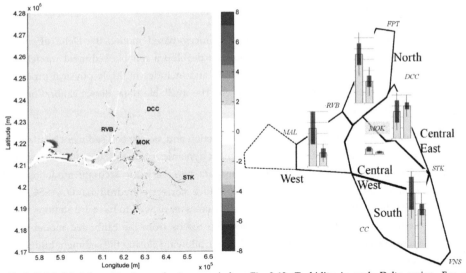

Fig 2-11: Modeled deposition in mm for 1 year period

Fig 2-12: Turbidity in each Delta region. For each region, the left bars indicate wet season and the right bars the dry season. The light gray bars indicate the mean turbidity over the region, the darker bars the spatial deviation and the lines the daily deviation. Each horizontal line represents 10 NTU.

2.4.6 Data input discussion

As a well-surveyed area that now has a complex process-based model, the Delta offers the opportunity to test how much data it is necessary to develop a reliable sediment model. The model supports high temporal and spatial resolution and includes multiple physical processes such as bottom friction, sedimentation, and erosion. The available data allows calibration and validation of model results.

As presented above, with simple settings of 1 mud fraction and simple bed sediment availability the model is capable of representing the main sediment dynamics processes, the peak timing, and duration, and results in a sediment budget. The data necessary for accurate modeling and forecasting is fine resolution bathymetry to reproduce correctly hydrodynamics, SSC, and discharge at the inflow and outflow boundaries. It is necessary as well to have 1-2 stations in the domain in order to properly calibrate the model. The results from the calibrated model using these few data can be extrapolated to the entire domain, allowing closing the sediment budget for the whole system.

The 2D model results output are available in high temporal (~hours) and spatial (~20 meters) resolution, and the modeled water quality parameters can be used in other models or for descriptive purposes. With limited input data, we can come to a detailed system description with considerable forecast capacity, expanding the applicability of this work to less-measured estuaries.

2.5 Conclusions

In this work, we make a step towards understanding and simulating sediment dynamics from source to sink in a complex estuary. This work shows that it is possible to reproduce the main system sediment dynamics as well as construct an accurate, detailed budget for complex areas such as the Delta using a 2D process-based numerical model coupled with a water quality model.

Overall, the model reproduces the SSC peaks and event timing and duration (wet season) as well as the low concentration in dry season throughout the Delta, except at Mallard where the water column is stratified due to salt intrusion. Stratification issues are not solved in a 2D model. For this reason, we are working on a 3D model in order to include the Bay area, leading to a unique source to sink model.

The Delta has many observation stations. However, this work shows that the substantial sediment is exported trough the pumping stations (100Kt yr^{-1}) at the Southern Delta where no data in SSC is available. This sediment export needs further investigation since it is possible that has been deposited in the channels before the pumps.

We show that with simple sediment settings of one fraction at the input boundary and a simple distribution of bed sediment availability, it is possible to reproduce seasonal variations as well as construct a yearly sediment budget with more than 90% accuracy when compared with a data derived budget. It also shows that it is extremely important to have discharge and SSC measurements at least in the input boundaries and close to the system output in order to be able to calibrate the model settings applied for hydrodynamics and suspended sediment. This methodology now can be applied in less-measured estuaries.

Sediment is a key factor in the water quality and ecology of an estuary. The Delft3D FM software allows direct coupling to water quality, sediment transport, and habitat modeling. Our work provides the basis for a chain of models, which goes from the hydrodynamics to suspended sediment, to phytoplankton, to fish, clams, and marshes. The turbidity and deposition pattern analysis may guide ecologists in future works to define areas of interest and/or venerable areas to study, as well as guide data collecting efforts. The present model opens the possibility for forecast and operational modeling. Forecasting the time frame of high levels of SSC (turbidity) allows planning of measurements campaigns for ecologists, as well as the possibility of tracking potentially contaminated sediment and be able to make a contingency plan as well as temporary barriers and pumping operations.

The Sacramento-San Joaquin Delta is a typical case of a highly impacted estuary. Being able to simulate numerically and determine sediment transport, budget and turbidity levels in this type of environment open possibilities to better informed political, ecological and management decisions including how to respond to climate change and sea level rise. This type of model is an important management tool that is applicable to other impacted estuaries worldwide.

Appendix 2-A: Hydrodynamic Calibration

The hydrodynamic calibration was carried out for 3-month high river flow conditions (December 16, 1999, until March 16, 2000) and a 3 month period of low river flow conditions (July 16, 2001, until October 16, 2001). All data is in NAVD88 (vertical datum), UTM 10 (horizontal datum) and GMT (time reference).

Hourly measured water levels at Point Reyes (tidesandcurrents.noaa.gov/) were used as a seaward boundary condition. Landward boundary conditions for the Sacramento River were obtained from daily measured river flow data at Freeport (FPT) and for the San Joaquin River near Vernalis (VNS) (cdec.water.ca.gov/). The inflow from the Yolo Bypass was approximated by curve fitting data from Qyolo and Qrsac.)

Measured data for the Bay area were obtained from tidesandcurrents.noaa.gov/, for part of the Delta from the California Data Exchange Centre cdec.water.ca.gov/ and for stations with numbers from direct contact with the Department of Water Resources (DWR).

Calibration was carried out by systematically varying the value of the Manning's coefficient for different sub-areas of the Bay-Delta system. The calibration data analysis includes (local and time varying) influence of air pressure and the wind in the definition of the boundary condition as well as in the calibration data inside the modeling domain. These may account for (part of) the error between measurements and modeling results. Also, the NAVD88 reference is not known for all measurement stations, although tidal water fluctuations may be modeled properly. To avoid these problems, a better method to assess the model performance is to focus on water level amplitude and phasing of the different tidal constituents. Boundary conditions, calibration data and model results are thus decomposed by Fourier transformation into tidal components which are then compared. By far, the main tidal constituents at (GGT) are O1, K1, N2, M2 and S2, with M2 being the largest. The model represents their values quite well. The difference in amplitude is 1.3 % for M2, up to 14% for O1, but the phasing shows a maximum of only 3% (O1).

Figure 2-A1 gives calibration results for the high and low river flow. The largest (extreme) deviations are explained by the fact that the measured water levels did not have a known reference to NAVD88 (http://www.D-Flow-baydelta.org/).

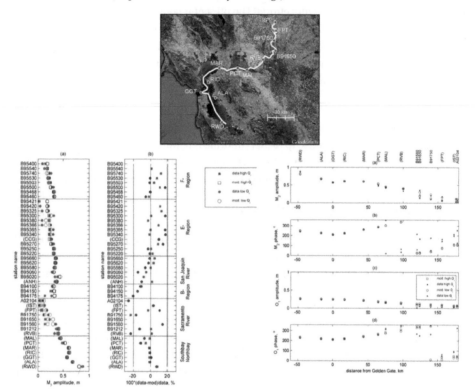

Figure 2-A1: Hydrodynamic calibration example.

Appendix 2-B: SSC Calibration

All stations clearly reproduce SSC peaks during high river flow periods and lower concentrations during the remainder of the year (apart from MAL during the July-August period). The good representation of the peak timing means that the main Delta event is reproduced by the model as well as the periods of Delta clearance. These two periods are critical for ecological models, and a good representation generates robust input to ecological models. A closer look at Fig 2-4 reveals differences between model results and data. These differences are discussed station by station in this appendix.

At RVB, SSC levels are directly proportional to Sacramento River discharge (Fig 2-B1), and that the model properly represents the water discharge peak intensity and duration. However, in the model, the first peak, which occurs in October, remobilizes sediment faster than observed in the data. Analyzing the raw data, it is possible to observe a trend of SSC increase which the model overestimates. A probable explanation lies in the initial sediment composition of the bed. Defining the bottom sediment composition does not account for consolidation processes; so the first peak comes after the dry season when the mud in the banks has consolidated. In the simulation case, when river discharge increases, it remobilizes non-consolidated bottom/bank sediment causing an earlier peak than in the data. Similar behavior is observed at STK in December. Sediment trapped in sub-aquatic vegetation and marshes could be another explanation for the slower increase of the first peak as the model discharges for both stations agree with data (Fig 2-4).

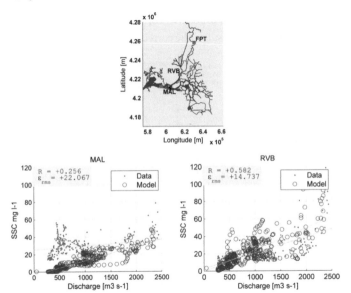

Figure 2-B1: Scatter plot Discharge versus SSC. Showing on the left (MAL) for MAL station and on the right hand (RVB) side RVB station. The red dots represent the Data and the blue model results.

Another difference between the data and the model results at RVB is the peak in May (second rectangle, (Fig 2-B2), which is not observed in the data. SSC level at the RVB station is directly proportional to water discharge in FPT (Fig 2-B1, RVB). The May peak is observed in FPT and so should have been transported towards RVB just as the two preceding peaks. However, the data set does not reproduce this peak. One of the possible explanations is an error in measurements since it comes after a major event and the equipment might be damaged. Other explanations could be a different composition of the suspended sediment properties and/or flocculation.

The model underestimates the first and second SSC peaks at MOK. However, the measured SSC signal is not consistent with the local water discharge signal. First, we checked that modeled water discharge is reproducing the local conditions, where data is available from mid-February onwards. The last peak in Figure 2-4 (mid-March) shows that water discharge, in situ and modeled SSC have the same range of variation. Therefore, the SSC levels are proportional to the local water discharge. Earlier, the January SSC data peak is much higher than the water discharge and the SSC level calculated in the model. The same happens in mid-February when no water discharge peak is observed, but there is a peak in the SSC data. Again the peaks in SSC could be caused by an error in the measurements or local, diffuse input of sediments such as from local farm wastewater or biological activity remobilizing the substrate.

The model represents the wet season SSC peaks well at MAL; however, during the three drier periods of the year the model underestimates SSC levels (Fig 2-B2 B). From the scatter plots of water discharge versus SSC (Fig 2-B1), it is possible to explain the weaker performance of the model during low river flow at MAL. These graphs represent river water discharge in FPT lagged by 2 days to SSC in RVB and MAL. Several time lags were tested, as MAL does not have a reasonable correlation with any of the time lags; it is presented here with the same time lag like the one for RVB. RVB station reflects a positive correlation between river discharge and SSC-derived from in situ data and model results. The correlation coefficient (R) at RVB is 0.58.

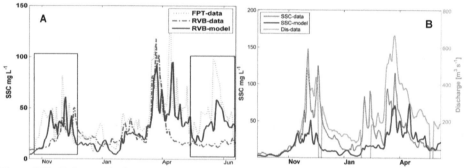

Figure 2-B2: A) Comparison between SSC levels in RVB station in situ data (dashed red) and the model result (solid blue) and FPT station (dotted green). B) Water discharge (model) and SSC level (data and model) in MOK station.

At the MAL station R=0.26, showing that there is not a strong correlation between river discharge and SSC levels. The low correlation is due to high SSC level during low water discharge periods, when the model underestimates SSC levels. Under low river discharges conditions, salt water intrudes into Suisun Bay leading to considerable stratification between fresh and salt water and shifting of the ETM landward (http://sfbay.wr.usgs.gov/access/wqdata/) (Brennan et al., 2002). In order to better model SSC levels for these conditions, a 3D model is needed at MAL. With this results, we are still able to calculate sediment export, since most of the sediment export occur in the wet period (McKee et al., 2006), when the model accurately reproduces measured SSC levels.

Acknowledgements

The research is part of the US Geological Survey CASCaDE climate change project (CASCaDE contribution 60). The authors acknowledge the US Geological Survey Priority Ecosystem Studies and CALFED for making this research financially possible. The data used in this work is freely available on the USGS website (nwis.waterdata.usgs.gov). The model applied in this work will be freely available from http://www.D-Flow-baydelta.org/.

3

SUSPENDED SEDIMENT DYNAMICS IN A TIDAL CHANNEL NETWORK UNDER PEAK RIVER FLOW

Peak river flows transport fine sediment, nutrients, and contaminants that may deposit in the network during or after the flush. This study explores the importance of peak river flows on sediment dynamics with special emphasis on channel network configurations. The Sacramento-San Joaquin Delta, which is connected to San Francisco Bay (California, USA), motivates this study and is used as a validation case. Besides data analysis of observations, we applied a calibrated process-based model (Delft3D FM) to explore and analyze high-resolution (~100 m, ~ 1 hour) dynamics.

Peak river flows supply the vast majority of sediment into the system. Data analysis of 6 peak flows (between 2012 and 2014) shows that on average 40% of the input sediment in the system is trapped, and that trapping efficiency depends on timing and magnitude of river flows. The model has 90% accuracy reproducing these trapping efficiencies. Modeled deposition patterns develop as the result of peak river flows after which, during low river flow conditions, tidal currents are not able to significantly redistribute deposited sediment. Deposition is quite local and mainly takes place at a deep junction. Tidal movement is important for sediment resuspension, but river induced, residual tide currents are responsible for redistributing the sediment towards the river banks and to the Bay.

We applied the same forcing for 4 different channel configurations ranging from a full Delta network to a schematization of the main river. A higher degree of network schematization leads to higher peak sediment export downstream to the Bay. However, the area of sedimentation is similar for all the configurations because it is mostly driven by the geometry and bathymetry.

This chapter is based on:
Achete, F. M.; van der Wegen, M.; Roelvink, D., Jaffe, B.: Suspended Sediment Dynamics in a tidal channel network under Peak River Flow, Ocean Dynamics DOI: 10.1007/s10236-016-0944-0

3.1 Introduction

Estuaries connect land and sea, host population and industries and are subjected to landward and seaward pressures. An important agent in this system is sediment that carries nutrients and contaminants (Bergamaschi et al., 2001; Bertrand-Krajewski et al., 1998; Deletic, 1998; Moskalski et al., 2013). Sediment is also responsible for nourishing marshes, mangroves, and beaches. Marsh restoration projects have become popular in the last years to protect inland from sea level rise. However, marshes need fine sediment to be able to grow and adjust to sea level rise (Boumans et al., 2002; Callaway et al., 2011; Fagherazzi et al., 2012; Prescott and Tsanis, 1997).

Flow velocity generates sediment transport, and spatial transport gradients cause erosion and deposition. In addition, in estuaries residual flow is caused by river discharge, tidal asymmetry, Stokes' drift, horizontal circulation, winds and channel geometry (Aubrey, 2013; Dronkers, 1986; Friedrichs and Aubrey, 1988; Guo et al., 2014; Hoitink et al., 2003; Rijn, 2011; van der Wegen and Roelvink, 2012; van der Wegen et al., 2008).

Mediterranean climate estuaries are highly seasonal with a short wet season (about 3 to 4 months) and a long dry season. Due to heavy rain and/or snowmelt, during the wet season the estuary receives peak river discharges that flush the system carrying sediment and pollutants to the estuaries (Bergamaschi et al., 2001; Obermann et al., 2009). In the San Francisco Estuary, the wet season dominates the sediment dynamics and defines the residual transport as river dominant though the high river flows may be present for several days only (McKee et al., 2006).

Channel geometry may have a considerable impact on sediment transport. Still, the geometry's influence on sediment erosion and deposition patterns is not fully understood. Anabranching is a river classification for multiple channels separated by vegetated islands. These networks include main channels, secondary channels, and connecting channels. It can be found in many environment types from high latitudes to subtropical and semi-arid (Nanson and Knighton, 1996), some examples can be found in San Francisco (Kimmerer, 2004), Texas (Phillips, 2014), Mekong (Meshkova and Carling, 2012), Amazon (Gallo and Vinzon, 2005), Guinea, (Capo et al., 2009) and China (Hu et al., 2011). In some cases, channels can migrate although they tend to be more stable in a cohesive sediment environment or in the presence of bedrocks and manmade levees.

Process-based numerical models are particularly suitable to explore scenarios of different forcing such as changing the geometry and boundary conditions. Although numerical models, because of the recent increase in computational power, now can consider more processes, longer time frames, and larger spatial scales, many studies seek to simplify the problem in order to understand the inherent estuarine processes (Ganju and Schoellhamer, 2009; Guo et al., 2014; Savenije, 2015; Townend, 2012). Simplifications such as estuarine numbers, non-dimensional numbers, and 1D model allow for a quick assessment of a study area without requiring high computational power.

This work aims to explore estuarine sediment dynamics under peak river flow conditions using a process-based model approach with special emphasis on the control of channel network configuration on seaward transport of sediment. A central question is how schematized a network can be and still accurately represent seaward sediment transport and deposition patterns. This work systematically decreases the complexity of a channel network estuary, defining several grids of decreasing channel complexity, to investigate the importance of river network in 1) sediment budget and 2) the deposition patterns in a channel network estuary.

3.2 Study Area

The Sacramento-San Joaquin Delta (Delta) consists of a collection of natural and man-made channel networks and leveed islands. The Sacramento River and the San-Joaquin River are the main tributaries (providing 90% of the freshwater discharge) followed by the Mokelumne River (Delta Atlas, 1995) (Fig 3-1). It is an important water convergence area of the U.S. West Coast collecting about half of the total California freshwater discharge. The Delta connects seawards to sub-embayments of San Francisco Bay near Mallard Island (MAL). In the Delta tidal currents change unidirectional river flow to bi-directional tidal flow with distinct ebb and flood currents. In the Delta, this region is located between Freeport (FPT) and Rio Vista (RVB) and varies with the river discharge, Q (Fig 3-1).

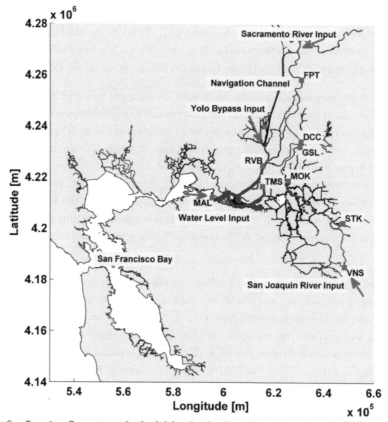

Figure 3-1: San Francisco Bay estuary. In the left-hand side, the embayments and in the right-hand side, the Sacramento-San Joaquin Delta location map. Sacramento and San Joaquin River are the main contributors of water and sediment discharge (blue). The boundary conditions are indicated by the red arrows.

The Mediterranean climate observed in this region defines the river discharge seasonality, where 70% of rainfall is concentrated between October and April (winter, wet) decreasing until the driest month in September (summer, dry) (Conomos et al., 1985). To facilitate the study and management decisions the year in the region is defined as water year (WY) and goes from October to September, e.g. WY 2012 ranges from October 2011 until September 2012. Studies show that Sacramento River accounts for 85% of the total sediment inflow to the Delta, while San Joaquin accounts for 13% (Wright and Schoellhamer, 2005). During the dry period, the average Delta discharge is 50-150 m^3s^{-1}, while during the wet period, the discharges range from 500-10000 m^3s^{-1} (Kimmerer, 2002). This discharge is observed in pulses with high sediment concentration (100-250 mg L-1) that flush the Delta system. In extremely wet years the discharge pulses can deliver in 1 day approximately 10% of the total sediment volume and 40% in 7 days (Ganju and Schoellhamer, 2006; McKee et al., 2006).

The geographic and seasonal flow concentration leads to several water issues related to agricultural use, habitat maintenance, and water export. On a yearly base, an average of 300 m^3s^{-1} of water is pumped from South Delta to Southern California (http://baydeltaoffice.water.ca.gov/sdb/sdip/features/ccf_diversions.cfm). The pumping rate is designed to keep the 2psu (salinity) line seaward of Mallard Island (MAL) avoiding salinity intrusion in the Delta.

3.3   Methodology

3.3.1 Model description

The numerical model applied in this work is Delft3D Flexible Mesh (Delft3D FM). Delft3D FM is a process-based unstructured grid model developed by Deltares (Deltares, 2014). It is a package for hydro- and morphodynamic simulation based on a finite volume approach solving shallow-water equations applying a Gaussian solver. The grid can be defined in terms of triangles, (curvilinear) quadrilaterals, pentagons, and hexagons, or any combination of these shapes. It is important to note that (orthogonal) quadrilaterals are the most computationally efficient cells. Kernkamp (2010) and the Delft3D FM manual (Deltares, 2014) describe in detail the grid aspects and the numerical solvers (Achete et al., 2015).

We assume that the main flow dynamics in the Delta are 2D (http://www.D-Flow-baydelta.org/). This implies that no salt-fresh water interactions occur in the Delta, and we assume that temperature differences do not govern flow characteristics. Delft3D FM generates hydrodynamic output for off-line coupling with the water quality model DELWAQ (Deltares, 2004). Off-line coupling enables faster calibration and sensitivity analysis. Delft3D FM generates time series of the following variables: cell link area; boundary definition; water flow through cell link; pointer file gives information concerning neighboring cells; cell surface; cell volume; and a shear stress file, which is parameterized in Delft3D FM using a Manning's n formulation. Given a network of water levels and flow velocities (varying over time) DELWAQ can solve the advection-diffusion-reaction equation for a wide range of substances including fine sediment. DELWAQ solves sediment source and sink terms by applying the Krone-Parteniades formulation for cohesive sediment transport (Ariathurai and Arulanandan, 1978; Krone, 1962) (Eq.3-1, Eq. 3-2).

$$D = w_s * c * (1 - \tau_b/\tau_d) \tag{3-1}$$
$$E = M * (\tau_b/\tau_e - 1) \tag{3-2}$$

Where; D Deposition flux of suspended matter (mg m^{-2}s^{-1}), w_s settling velocity of suspended matter (ms-1), c concentration of suspended matter near the bed (mg m^{-3}), τ_b bottom shear stress (Pa) τ_d critical shear stress for deposition (Pa), E erosion rate (mg m^{-2} s^{-1}), M first order erosion rate (mg m^{-2}s^{-1}), τ_e critical shear stress for erosion (Pa).

Following Winterwerp (2006) we assume that deposition takes place regardless of the prevailing bed shear stress. τ_d is thus considered much larger than τ_b and the second term in equation (Eq3-1) is neglected.

To study the influence of the channel network configuration on the deposition patterns we created 4 different networks (Fig 3-2). The average cell size is 100x30 meter for the main rivers decreasing to 20x5m in the creeks. The first grid is of the entire Delta (grid "delta") and comprises the Sacramento River from Freeport (FPT) and San Joaquin River from Vernalis (VNS) up to Mallard Island (MAL), as well all smaller rivers, channels, creeks and flooded island (Fig 3-1, Fig 3-2a). At FPT and VNS river flow is unidirectional at high discharge, tides are unimportant, and MAL is the Delta seaward limit. The second grid (grid "2 rivers", Fig 3-2b) keeps the 2 main rivers and the connections between them including the Delta Cross Channel (DCC), Georgiana Slough (GSL) and Threemile Slough (TMS). The third grid (grid "sacra extension", Fig 3-2c) comprises only the Sacramento River and the north branch that connects with the Yolo Bypass. The Yolo Bypass is a flood bypass to protect Sacramento from flooding; it diverts Sacramento River water at Freemont weir upstream from FPT and connects again with the Sacramento River at Liberty Island. The Yolo is currently a wetland wildlife area. The fourth (grid "sacra", Fig 3-2d) grid represents only the Sacramento River from FPT to MAL.

Hydrodynamics and sediment transport is simulated for the 4 grids. First, we ran Delft3D FM to calculate water level, velocities, cell volume and shear stresses, and then DELWAQ calculates SSC levels. The "delta" grid was the standard for calibrating, at RVB station, water levels, discharges and SSC levels against in-situ data. From the SSC levels and discharge, it is possible to calculate sediment flux and its gradients, which result in deposition or erosion.

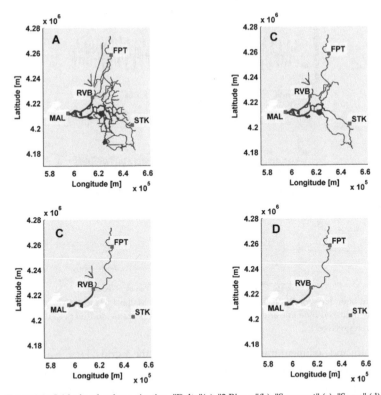

Figure 3-2: Grid of each schematization, "Delta"(a), "2 Rivers"(b); "Sacra ext" (c); "Sacra" (d).

3.3.2 Initial and boundary conditions

The bathymetry is based on different data sources including bathymetric soundings collected from 1933 to 2012 and Lidar data collected from 2007 to 2010. The initial bathymetry has a 10m grid resolution, which is based on an earlier grid (Foxgrover et al., http://sfbay.wr.usgs.gov/sediment/delta/), modified to include new data by Wang and Ateljevich (http://baydeltaoffice.water.ca.gov/modeling/deltamodeling/modelingdata/DEM.cfm) and further refined.

The number of open boundaries conditions varies depending on the grid. For all grids, the seaward boundary is defined as a water level derived from MAL station (http://cdec.water.ca.gov/) and the landward boundary of Sacramento River water discharge and sediment concentration derived from FPT station. The grids "delta", "2 rivers" and "sacra extension" include the Yolo Bypass and San Joaquin River (VNS) landward boundaries for water discharge and suspended sediment concentration (SSC) (Fig 3-3). Rio Vista (RVB) is the station for hydrodynamic (water level and discharge) and sediment calibration (SSC) for the grid "delta".

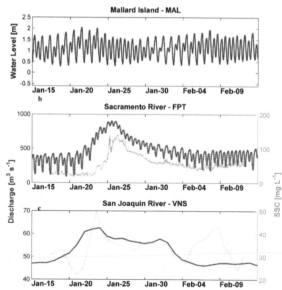

Figure 3-3: Input boundary condition. Top panel (a) water level at Mallard Island (MAL), the following 2 panels show discharge in dashed blue line and SSC in solid green line for Sacramento River at FPT (b) and San Joaquin River at VNS (c) respectively.

Hydrodynamics and sediment transport is simulated for the 4 grids. First, we ran Delft3D FM to calculate water level, velocities, cell volume and shear stresses, and then DELWAQ calculates SSC levels. The "delta" grid was the standard for calibrating, at RVB station, water levels, discharges and SSC levels against in-situ data. From the SSC levels and discharge, it is possible to calculate sediment flux and its gradients, which result in deposition or erosion.

All landward boundaries and calibration data were obtained from (http://maps.waterdata.usgs.gov/). The continuous monitoring program of the USGS publishes 15 minute SSC data from optical backscatter sensor (OBS) calibrated with bottle samples approximately monthly (Wright and Schoellhamer, 2005). Initial SSC was set at 20mg L⁻¹ over the entire domain. There was no sediment available on the initial bed and no bed level change was allowed.

3.3.3 Calculation of sediment discharge and bed level change

The analysis of flow and suspended sediment observations at river stations and model outputs use a methodology similar to the one presented by Downing-Kunz and Schoellhamer (2015). The upstream station FPT has unidirectional flow and represents the river station, the downstream station RVB is tidal influenced and the last station without salinity influence. Here we analyze deposition/erosion calculating mass storage (ΔS), which is the difference between time integrated suspended sediment discharge upstream (SSD_{FPT}) and downstream (SSD_{RVB}), $\Delta S = \int_0^N SSD_{FPT} - \int_0^N SSD_{RVB}$, where N is the total time interval. Positive mass storage indicates sediment deposition and negative sediment erosion. Sediment deposition indicates that sediment is trapped in an area, so we can calculate the trapping efficiency by $\Psi = \frac{\Delta S}{\int_0^N SSD_{FPT}}$. The bathymetric area (A) between FPT and RVB is about 44 km2 and bulk density (ϱ) 1300kg m-3. From this information it is possible to calculate deposition height $h = \frac{\Delta S}{\rho * A}$. For the data the deposition is considered homogeneous distributed over the area, but the model gives more detailed information of amount of sediment per cell allowing a better estimation of sedimentation height and patterns.

As we are interested in the residual transports, the results are also analyzed in terms of tide averaged results. Applying a Butterworth filter with cutoff frequency of 1/30 h^{-1}(low pass filter) all variations with a frequency higher than approximately 2 days are filtered (Scully and Friedrichs, 2007; Wright and Schoellhamer, 2005).

3.3.4 Model calibration and dynamics

The "delta" grid represents the most realistic case since it considers all Delta channels and respective bathymetry. For this reason, the "delta" grid was calibrated against data for water level, water discharges and SSC at RVB station (Fig 3-2). RVB has available data for the aforementioned variables, experiences tidal currents and has no influence of salinity intrusion, therefore, no stratification. The calibration settings for Manning friction coefficient (spatially variable, 0.017-0.03), eddy viscosity (1), and eddy diffusivity (1) derived from the "delta" grid calibration was then applied to all other grids. The "delta" grid is the only case where it is possible to calibrate water level and discharge, and so tune the parameters.

The model reproduces the mixed meso-tidal dynamics observed in data. At RVB, the neap-spring tidal cycle is ~1.2m during neap and ~1.6m during spring. The diurnal inequality during neap is ~0.2m and during spring ~0.5m. The "delta" grid reproduces discharges phase and amplitude with a correlation of 98% (R - correlation coefficient) with the data. During spring tide, discharges range from -1700 m³s⁻¹ (landward) to 2300 m³s⁻¹ (seaward) (Fig 3-4). Filtering the tidal variation, it is possible to identify the peak river event between January 21st and February 4th when net discharge increases from 200 m³s⁻¹ to 900 m³s⁻¹.

As for the hydrodynamic forcing the "delta" grid was calibrated for SSC (Fig 3-4). The calibration parameters were fall velocity (0.4 mm s⁻¹), critical shear stress (0.20Pa) and erosion coefficient (10-4kg m³s⁻¹). Net sediment transport is related to residual currents, and it is possible to estimate sediment deposition and erosion by spatial gradients in SSD (ΔS). The aforementioned results give the confidence to proceed to the result analysis of the channel schematizations.

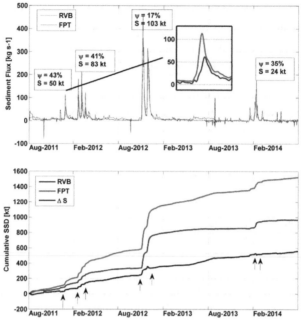

Figure 3-4: Model (blue) versus data (dashed red) at RVB station a) water discharge, b) suspended sediment concentration and c) suspended sediment discharge for an entire month (15-Jan until 15-Feb 2012).

3.4 Results

3.4.1 Mass storage between Sacramento River stations

This section presents a brief analysis of mass storage between FPT and RVB using observations at river stations. To calculate mass storage, it is necessary to have observations of discharge and SSC for two stations. This data is available for FPT and RVB for water years (WY) 2012-2014 (nwis.waterdata.usgs.gov).

We are interested in the processes related to river pulses. During the period of analysis 4 events occurred, 2 of which consisted of more than one peak. Observations in Corte Madera Creek (Downing-Kunz and Schoellhamer, 2015) show that sediment input fluvial contribution during river pulses is trapped in the river due to river tide interaction. In this work we analyze whether the Delta has the same behavior as Corte Madera, regarding sediment trapping in Sacramento River, and calculate the deposition rates related to peak discharge. USGS provides daily statistics for the last 66 years for Sacramento River at FPT (Fig 3-5a). From the WYs analyzed, WY 2012 has average discharge with 4 peaks distributed during the 4 months of the wet season. Comparing only the highest discharge peak, the peaks of WY 2012 are lower than the average peak discharge (1800 m³s⁻¹) (Fig 3-5a). WY 2013 shows river pulses earlier and higher than average. In the WY 2013, the 2 peaks are in November and December while normally the peaks are between January and April. Although it has a short wet season, the mean discharge is higher than average wet period discharge. In contrast, WY 2014 is an extremely dry year that has a total river flow that is 64% less than average.

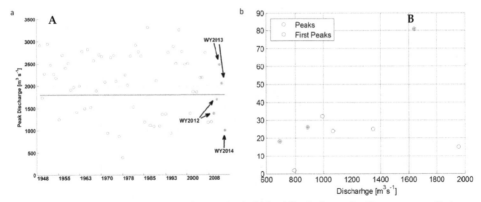

Figure 3-5: A) Peak discharge from 1949 until 2014. The dashed red line indicates the 66-year average discharge. In B) the peak discharges for the first peak flows for WY 2013 and 2014 are highlighted in red.

The first peak after a dry season may transport more sediment than similar peaks over the year (Goodwin and Denton, 1991). In general, this behavior is observed in the data set analyzed (Table 3-1). For first flow peaks the ratio SSC/Q, where Q is water discharge, is higher than for following peaks of similar magnitude for WY 2013 and 2014 agreeing with (Goodwin and Denton, 1991) (Fig 3-5b). For WY 2012 the first peak flow is much smaller than subsequent peak flows. Peak 2 and 4 provide smaller SSC/Q ratios than the first peak, but peak 3 has a larger SSC/Q ratio (Table 3-1). This may be due a dependency of suspended sediment discharge, SSD, on flow magnitude. Both timing and magnitude will have an effect on the SSC/Q ratio. A larger data set may reveal a more significant trend between SSC and river flow magnitude.

Table 3-1: Observed Freeport peak river discharge, the ratio of observed suspended sediment concentration to river discharge and calculations of mass storage, trapping efficiency, and average deposition from spatial gradients in suspended sediment discharge for each peak event of the 3 water years studied. First flow peak of the water year indicated by gray background.

Peak	Peak Discharge (Q m³s⁻¹)	SSC/Q At FPT	Mass Storage (ΔS kt)	Trapping Efficiency (Ψ %)	Deposition (mm)
2012 - 1	891	0.16	50	43	0.006
2012 - 2	1068	0.13	26	45	0.004
2012 - 3	1351	0.13	25	40	0.004
2012 - 4	996	0.23	32	38	0.006
2013 - 1	1639	0.22	86	29	0.01
2013 - 2	1957	0.10	17	6	0.003
2014 - 1	694	0.15	20	69	0.004
2014 - 2	796	0.12	4	2	0.0006
Total or Average			260 / 32	34	0.038 / 0.005

Sediment passing FPT flows downstream to RVB analyzing the 3 years of SSD. For both stations, we observe an average decrease in SSD of 34% (trapping efficiency, Ψ) or a positive mass storage (ΔS) of 559kt (Table 3-1, Fig 3-6). Of the total ΔS, 299 Kt is deposited during the dry season deposition while 260kt is deposited during the wet season. Notice that the wet season corresponds to only 6 months out of the 36 months analyzed. The average monthly deposition during dry months is about 10 Kt or about one-quarter of the 43kt deposited during wet months.

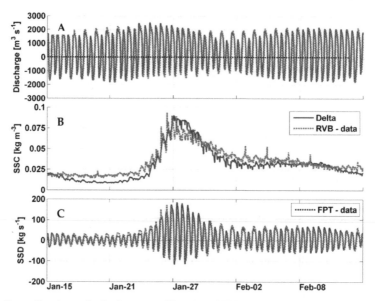

Figure 3-6: Sediment flux time series for Sacramento River, in red FPT (upstream) and in blue RVB (downstream). In the panel is a zoom showing the decay of SSD between the two stations. Cumulative sediment discharge, upstream at FPT (red), downstream at RVB (blue) and the mass storage (ΔS), which is the difference between SSD at FPT and SSD RVB (black). The arrows indicate the peaks.

At the peak flow timescale, trapping efficiency varies depending on multiple factors such as discharge, available bed sediment and tidal excursion (Downing-Kunz and Schoellhamer, 2015). During the 3 years analyzed the peaks have great variability resulting in ΔS and Ψ with means and standard deviations of 33kt +/- 25kt, and 34% +/- 22%, respectively (Table 3-1, Fig 3-6). Sedimentation rates are highest during the peak events, rapidly decreasing after the event (Fig 3-6b).

Also, there is a negative ΔS (net erosion) just before the second SSD peak, e.g. February 2012, December 2012 and February 2014 (Fig 3-6b). A possible explanation is that the first peak brings sediment which deposits on the bed while it has only limited time to consolidate before the second peak arrives. When the next peak discharge arrives, the deposited sediment from the previous peak is washed away. This phenomenon is mainly observed when the second peak occurs just after the first peak of the year, as in WY 2013 and WY 2014. WY 2013 second peak has a very low Ψ (2%), probably due to a combination long preceding dry season, and a very low first peak which does not reset the system leaving most of the sediment available to be washed away in the second peak.

In situ data does not provide insight into sediment distribution. In order to translate ΔS into sedimentation height (mm) we consider evenly distributed sediment over the area with a density of 1300 kg m-3. The calculated heights range from 0.005mm to 0.01mm of deposition for the different water years. During the analyzed period, the yearly average deposited sediment height for peak events was 0.005mm and the total for the 3 years 0.038mm (Table 3-1).

Data and literature analyzes show that sediment arrives in the estuary in peak events. Sediment deposits in the upper estuary during and after the peak events. The upper estuary is where tidal currents change the unidirectional river flow to bidirectional flow, experiencing ebb and flood currents. This behavior is expected in event driven estuaries, where the peak river events dominate the sediment dynamics. In the Delta, the limit of the estuary is located between FPT and RVB and varies with river discharge. The seawards boundary of the Delta is at Mallard Island, where the Delta connects to San Francisco Bay.

3.4.2 Hydrodynamics

We compared model results for the full grid "delta" to the schematized grids to explore the response of the tidal discharge and velocity to the channel schematization at RVB. In "sacra" and "sacra ext" the reduction of the tidal prism is too large; therefore, the specified tidal range at the boundary is probably overestimated.

The tidal discharge and velocity amplitude in "2 rivers" and "sacra ext" are similar to the "delta" (Fig 3-7a, b). The "2 rivers" has a lower filtered discharge at RVB (Fig 3-7e), due to a higher southward water flow through Georgiana Slough (GSL). While "delta" has a peak of 140m³s⁻¹ flowing through GSL, "2 rivers" has 210m³s⁻¹ flowing through GSL. The DCC and TMS are two other connections between Sacramento River and San Joaquin River. DCC is closed in the model during most of the peak discharge due to Delta barrier operation to fish conservancy plans. TMS net discharge is on average 50m³s⁻¹ and is not affected by the Sacramento River peak (Achete et al., 2015).

The "delta" grid discharge flows into Eastern Delta channels at about 0-5 m³s⁻¹, which are very low compared to the Sacramento and San Joaquin River discharges (~500m³s⁻¹, 80m³s⁻¹, respectively). As a result, the difference in residual current between "delta" and "2 rivers" grids is mainly due to GSL.

"Sacra ext" has a higher filtered discharge than "delta" and "2 rivers" since the water has no other path to flow. The grid "sacra" has 1/3 of the discharge and velocity amplitude of the other grids (Fig 3-7a, b); even though the tidal filtered results show a good agreement for all the grids (Fig 3-7, e, f). This suggests that that the Sacramento River flows govern the net discharge, although maximum tidal flows are almost triple than the net flows.

3.4.1 Suspended sediment discharge

The schematizations add variability to SSC (Fig 3-7c), as well as increase the value during peak discharge events (Fig 3-7g). SSC and consequently SSD for "2 rivers", "sacra ext" and "sacra" grids have an earlier peak compared to the "delta" grid (Fig 3-7g, h) and in the more schematized grids SSD is up to 50% larger than in "delta"(Fig 3-7 h).

Before and after the peak discharge, SSC in the schematized grids are 40% less than when using the "delta" grid, with the exception of "sacra ext", which has similar SSC to the "delta" grid.

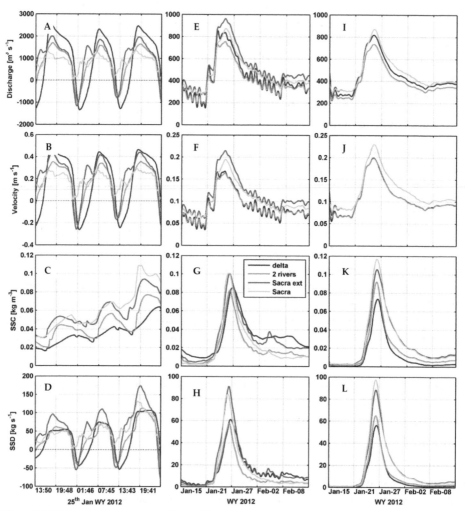

Figure 3-7: All the panels represent results at RVB station. Panels a-d are a zoom of the entire time series during high flow, positive discharge and velocity are seaward oriented. Panel e-h are the tidal filtered results for an entire month (15-Jan until 15-Feb 2012) and panels i to l the results from the no tide case. Panel a, e and i show water discharge, b, f and j velocity, c, g and k SSC; d, h and l SSD. Lines for Sacra ext coincide with "sacra" in (i and j).

3.5 Discussion

3.5.1 Hydrodynamics

The grids schematizations can be seen as a decrease in the tidal prism, which is the wet area times the difference between mean low water level (MLWL) and mean high water level (MHWL).

Changes in tidal prism lead to a change in velocities and discharge values (Horrevoets et al., 2004; Savenije, 2001).

The realistic case grid "delta" has a wet area of 200 km², the "2 rivers" the area is 134 km², "sacra ext" 56 km² and "sacra" 43 km². The difference between MLWL and MHWL increases from "delta" (1.5 m) to "sacra" (2.1m). Comparing the changes in the tidal prism, we observe that "2 rivers" (0.26 km³) represents 87% of "delta" prism (0.29 km³). The change is more dramatic to "sacra ext"(0.12 km³) that represents 39% of the initial tidal prism, decreasing to 30% to "sacra"(0.09 km³).

A higher level of channel schematization leads to a lower tidal prism decreasing the tidal discharge (Fig 3-7a). The change in the tidal prism of "sacra" is so extreme that discharge is always seaward. The level of schematization does not significantly affect the phasing of tidal discharge (Fig 3-7a).

3.5.2 Suspended sediment discharge

Analyzing SSD brings us back to the work by (Downing-Kunz and Schoellhamer, 2015). The main reason that SSD are different for different grids is that SSC is very sensitive to prevailing velocities. Tidal velocities are high in "delta" and "2 rivers" due to a larger tidal prism. Yet, residual velocities are larger in the more schematized grids. Two main reasons for these higher residual velocities were identified. First the higher mean water (water setup) in the upper estuary (FPT region), that increases the river velocity, which is especially the case during high river discharge (Fig 3-8). Secondly, tidal currents are not high enough to suppress the velocity of the river signal (Fig 3-7, b). The higher velocities keep sediment in suspension for longer increasing SSD through RVB during peak events.

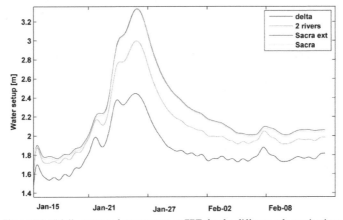

Figure 3-8: Tidally averaged water setup at FPT, for the different schematizations.

Like "delta", "2 rivers" represents the SSD peak quite well, albeit a couple of hours earlier than "delta". The difference in phasing is because the tidal velocity in "2 rivers" does not suppress the river signal as much as in "delta". Table 3-2 shows that "2 rivers" overestimates ΔS by 30% compared to "delta", and a 40% overestimation of the data. "2 rivers" has a trapping efficiency of

66%, the "delta" 47% and the data 43%. This implies that the small channels considered in the "delta" model are responsible for only 13% of the main sediment entrapment.

Table 3-2: modeled mass storage, trapping efficiency, and average deposition for each schematization run spanning WY2011.

Peak	Mass Storage (ΔS kt)	Trapping Efficiency (Ψ %)	Deposition (mm)
Data (2012-1)	50	43	0.006
Delta /no tide	55 / 87	47 / 75	0.03
2 rivers /no tide	71 / 81	66 / 69	0.02
Sacra ext/no tide	44 / 50	38 / 43	0.015
Sacra /no tide	62 / 49	53 / 42	0.015

The higher mass storage in "2 rivers" has 2 main reasons. The first one is the divergence of the peak discharge to GSL which traps more sediment than the Sacramento River. The second is the smaller tidal velocities in "2rivers" than in "delta" grid, so that, after the SSD peak, the sediment is more easily deposited.

The high velocities in `sacra ext` and `sacra` keep the input sediment from FPT in suspension for a longer time, leading to a higher SSD through RVB. The "delta" SSD peak is around 60 kg s^{-1} while in "sacra ext" and "sacra" it reaches ~85 kg s^{-1}, though it has the same phase as "delta" and "2 rivers".

"Sacra ext" is the intermediate case, so its dynamics is a mixture between the more complex grids ("delta" and "2 rivers") with pronounced tidal variation during the low river discharge and the simpler grid ("sacra") with high SSD. Even though the river velocity is suppressed by tidal currents, the peak discharge has no other path to go but to flow towards RVB, increasing the net velocity and exporting more sediment than the more complex networks. The overestimation of peak SSD and the tidal current in drier period keeping sediment in suspension leads to lower ΔS (44kt) consequently lower Ψ (38%).

"Sacra" grid overestimates the peak SSD due to high velocities during peak river discharge. However, after the SSD peak, the low tidal velocity does not re-suspend the deposited sediment resulting in a high ΔS of 62kt and Ψ of 53%, similar to "delta" (47%) and data (43%).

3.5.3 Deposition pattern

The deposition pattern is largely determined by the geometry and bathymetry of the Sacramento River. At the junctions with Yolo Bypass and the navigation channel, the Sacramento River course changes direction at the same time the depth increases from 2m to 12m so that velocities decrease inducing sedimentation.

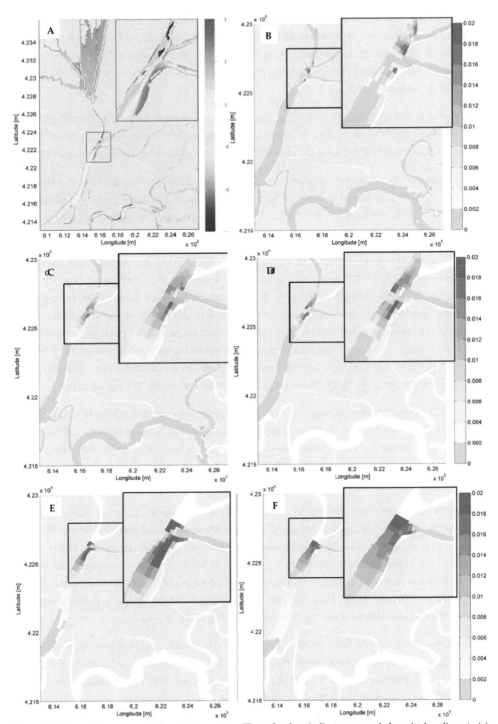

Figure 3-9: Deposition maps at the last time step. The color bar indicates mm of deposited sediment, (a) bathymetric data, (b) "delta", (c) "2 rivers", (d) "Sacra ext", (e) "Sacra" and (f)"delta no tides".

At the junction sedimentation is also observed in the data (Fig 3-9) as a 200 m long bar on the Eastern bank. The levees, represented as the black line in figure 3-9a, was primary bordering the marshes, so all the sediment westward from the levee was deposited after the construction. We stress that DELWAQ does not update bathymetry, so this value represents a virtual thickness of sediment deposited/eroded.

This region is just upstream from RVB, and it is the main area of deposition. "Delta" grid traps 55t, 10% more than the ΔS derived from calculations using observations (50t). However, there is just 4% of difference in Ψ between the model (47%) and calculations using observations (43%).

Townshead (2013) indicates considerable dredging to keep the navigation channel from Sacramento to Suisun Bay at a depth of 10 m; approximately 115,000m³ (~150t) of sediment is dredged in 2013. Approximately two-thirds of the dredged material are taken from the shipping channel and the other third at the Sacramento River main channel (~50t).

To analyze the temporal variation of the deposition we selected the regions with cumulative deposition greater than 0.02mm when averaged in space for the "delta" grid (Fig 3-10). The sedimentation can be divided into 3 periods. The first period goes from the beginning of the simulation until the peak discharge. Little sedimentation is observed until the middle of the peak discharge. The maximum sedimentation occurs at the arrival of the SSD peak, just after the peak water discharge, up to a week after the peak passage, when the deposition rate starts to decrease again. Deposition flux (gradient of sediment height line) is equal at the start and end of the simulation and is the same as the deposition rate observed during the dry period in Fig 3-6.

The calculations using observations of Q and SSC give an average of 0.005mm of deposition for each peak discharge event. This value is one order of magnitude lower than the model results because of the model we know where the deposition occurs and calculate the deposition height based on cells where deposition occurs while in the calculation using observations the sediment volume is spread over the entire river between the stations.

The "delta" grid reproduces the bar location, close to the east bank and scales, ~200m long and ~50m wide preserving the main channel (Fig 3-9b). We observe sedimentation north from the junction as well, in this case, the sedimentation has the same order of magnitude, but it is more spread in the channel than in the south case. Close to MOK station sedimentation in order of 0.005-0.02mm is observed. For small channels in the Eastern part of the Delta, sedimentation is on the order of 0.001-0.003mm. However, the sedimentation height is one order of magnitude smaller than sedimentation at the junction and at MOK (0.02) (Fig 3-9).

At RVB, the "2 rivers" grid has a lower SSD than the "delta" grid, implying more deposition. The higher deposition is linked to a higher discharge at GSL carrying more sediment to the Eastern Delta, which deposits a layer of ~0.005mm close to MOK station and in San Joaquin River (Fig 3-9c). The "2 rivers" deposition in the Sacramento River agrees with the "delta" grid (~0.03mm) of deposition height, 0.008-0.020mm, and location. As in "delta", "2 rivers" shows deposition north and south of the junction. However, the deposited sediment occupies a larger area and is more evenly distributed across the channel for about 1km in both directions. For the cases "delta" and

"2rivers," the sedimentation volumes in smaller channels are at least one order of magnitude lower than sedimentation volumes in the main rivers.

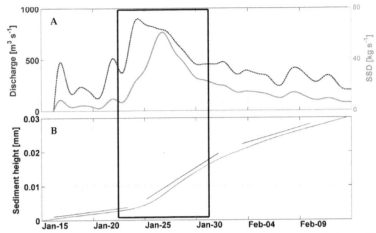

Figure 3-10: A) River discharge at RVB (dashed blue) and suspended sediment discharge data at RVB (green). B) mean sediment height variation in time; the gray lines indicate the sedimentation rate. The black rectangle highlight the peak period for WY2012.

Remarkably, the "sacra ext" (Fig 3-9d) and "2 rivers" lead to similar features in the deposited sediment at the junction as the full "delta" grid. The deposition concentrates at the junction Eastern bank similar to "delta" though it is not possible to distinguish a clear channel by the deposition pattern. The presence of more complex patterns in "sacra ext" can be explained by the higher velocities of tidal currents that stir bottom sediment increasing sediment remobilization.

Downscaling, "sacra" still has the main deposition area at the junction. As in "2 rivers", the deposition is evenly spread at the bottom, but it covers almost double the area and is distributed over about 2km (Fig 3-9e).

3.5.4 Tidal influence

As discussed in previous sections the estuary schematization reduces the tidal prism leading to smaller tidal currents, which are not strong enough to suppress the river discharge signal. Therefore, the residual currents mainly driven by river signal are higher in smaller grids (Fig 3-7 f). This difference in velocity is less noticeable in the net water discharge (Fig 3-7 e); but they are important for keeping fine sediment in suspension increasing the suspended sediment discharge (Fig 3-7g, h).

Tests with no tides at the seaward boundary were done to verify the impact of the tidal signal in the model for the different schematizations. For these simulations, the same grids, model settings for hydrodynamics and suspended sediment were used, and the only difference was the water level boundary at MAL, which was set at 0m elevation (mean sea level).

We observe a change in the shape of the river discharge for "delta" and "2 rivers", the 2 grids that produced results most affected by tides. The net discharge magnitude has not changed, since, as previously discussed it is river driven (Fig 3-7i). Figure 3-7 j (no tide) shows higher velocities than Figure 3-7f (tide case). In the no-tide case, for all the grids, both the velocities (Fig 3-7 j) and SSCs (Fig 3-7k) are in phase, corroborating the aforementioned discussion of tidal current suppressing the river signal velocity.

As there is no tide, the SSD is directly proportional to the number of places where sediment can be diverted from the Sacramento River (Fig 3-7 l). The "delta" and "2 rivers" have lower SSD than the other grids because there are more connections to the Eastern Delta where sediment can escape. As "delta" has more connections than "2 rivers" it diverts the most sediment (Fig 3-7 l). "Sacra ext" and "sacra" have the same SSD, since there are no secondary channels between FPT and RVB and no tide to suppress the flux.

During peak discharge the river dominates the dynamics, so the SSD peak magnitudes of the tide and no-tide cases are similar. Nevertheless, tidal currents are important for stirring sediment, modifying the SSD peak duration, and transport before and after the peak. Mass storage and trapping efficiency (Table 3-2) show the importance of the tides in reworking and exporting the sediment. For the "delta" case Ψ increases by 27% with no tides, and it is almost the double of the value observed in the data. As showed before the Ψ decreases with the increase of schematization; "sacra" grid Ψ is 42% with no tides, a quite similar value to the data (43%) but with a completely different behavior. "Sacra" experiences a high SSD during the peak but after the peak, virtually no sediment is exported, while in the grid "delta", the transport increase during the peak week, and it keeps a baseline transport of about 20kg s^{-1} during the rest of the month (Table 3-2).

The sedimentation pattern of the no-tide case agrees with previous findings of the tidal current remobilizing the deposited sediment at the junction. Figure 3-9 b shows deposition close to the river banks while Figure 3-9 f the deposition is spread over the river channel. At the no-tide case, all the grids have the same behavior, the main deposition area is at the junction, and it is evenly spread in across channel. At MOK, the deposition is shifted northward.

3.5.5 Simulating the second discharge peak

Runs were carried out with initially no sediment available on the bed. Erosion of the initial bed was thus also not possible. This condition may reflect bed conditions after a dry season during which there was only limited supply of sediment and where deposited sediment during the peak river flow of previous year had consolidated. However, it may be questioned whether sediment fluxes during a second peak originate from the bed (locally suspended) or are sediment load supplied by the peak discharge. To explore these dynamics we carried out a test considering 2 identical subsequent peaks with two sediment fractions of equal characteristics as if we were color-coding the sediment of each peak.

For this test DELWAQ reads the "delta" hydrodynamics results twice, reproducing 2 equal peak flows. It is possible to track sediment particles of each peak separately, IM1 for the first peak and IM2 for the second peak. The second period begins on 15th of February, and the peak discharge is 10 days later (25th of February). At this time, IM2 represents 10% of the total flux. The arrival of the new flow peak stirs deposited IM1 and transports it through RVB (Fig 3-11). The 2nd flow peak leads to higher SSC at the start of the peak and slightly higher SSC at the peak itself. It does not change the phasing between peak flow and peak SSC.

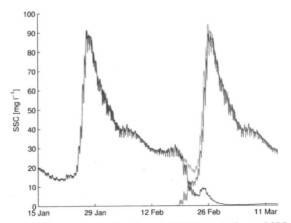

Figure 3-11: Two consecutive SSC peaks are passing through RVB. Green is the total SSC, in blue sediment from the first peak and in red from the 2 peak.

So 12 days from the start of the new 2nd period the peak SSD transporting IM2 arrives, and IM2 becomes the dominant fraction transported. Hence, before the second peak, the main source of sediment is the deposited sediment from previous peaks (IM1), arriving the peak SSD IM2 dominates the exported sediment.

The 2 peaks simulation does not reproduce the difference in trapping efficiency between the first and second peak as suggested by data analysis. In the simulation, both SSC peaks have the same magnitude. For the simulations, it was considered a simple parameter setting of one mud fraction, no initial sediment availability and no bottom layering (Achete et al., 2015). The lack of difference in trapping efficiency could be related mainly to the 1 sediment layer on the bed and no initial sediment availability as well as sediment stabilization by biota and consolidation.

3.5.6 Recommendations

Currently, the model does not include a morphologic feedback so the impact of changing bathymetry could not be assessed. In the short term (~years), it is not important for this system because the deposition rates are very small (~0.04mm, Table3-2). However, the inclusion of morphologic feedback might be relevant for other similar systems with higher sedimentation rates. For future work, the model results may be improved by considering multiple layers, with increasing critical shear stress as a parameterization of bed consolidation. The inclusion of spatial variation in colonization by vegetation and biota, which reduce erosion rates by stabilizing the bed, may also be a productive area of research that improves model results.

3.6 Conclusions

Peak river flows determine to a large extent the sediment dynamics in the Sacramento-San Joaquin Delta. Analysis of river station observations shows that trapping efficiency of the sediment entering the Delta during peak flows varies depending on the timing and magnitude of the river flows. Trapping efficiency is around 40%, although significant deviations (2% and 69%) are found as well. A more detailed analysis of these dependencies could and should be made when more data become available in future.

The process-based model reproduces trapping efficiencies quite well (47% compared with an observed 43% for the first peak flow event in WY2012). The model allows for a detailed analysis of sediment dynamics. Deposition patterns develop as the result of peak river flows after which, during low river flow conditions, the tidal currents are not able to significantly redistribute deposited sediment. Deposition is quite local and mainly takes place at the junction in the region where the Sacramento River, the deep water shipping channel and Yolo Bypass merge. This is probably a deep region subject to dredging to maintain shipping to Sacramento via the deep water shipping channel. We could not confirm this with data. The limited impact of tidal flows is confirmed by runs without a seaward tidal forcing showing similar hydrodynamics and sediment dynamics. However, no-tide runs result in an increase in trapping efficiencies because the tidal movement enhances sediment suspension. The model could not explain the lower trapping efficiency of second flow peaks as suggested by the data analysis. Possible reasons for this are processes not included in the model as bioturbation and consolidation. Future works considering bed layering and spatial distribution of shear stress could improve the differentiation between first and second peak.

More schematized networks under equal forcing lead to remarkably similar deposition patterns. Excluding smaller channels in the network decreased mass storage by about 15%. A higher level of schematization also leads to higher tide residual velocities, more sediment export, and less trapping efficiency. These results provide guidance for modeling of less measured estuaries where not all the small channels have bathymetry data, and still are able to calculate mass storage and trapping efficiency.

Acknowledgements

The research is part of the US Geological Survey CASCaDE climate change project (CASCaDE contribution 65). The authors acknowledge the US Geological Survey Priority Ecosystem Studies and CALFED for making this research financially possible. The data used in this work is freely available on the USGS website (nwis.waterdata.usgs.gov). The model applied in this work will be freely   available   from   http://www.D-Flow-baydelta.org/.

4

IMPACT OF A SUDDEN TIDAL PRISM INCREASE IN ESTUARINE SEDIMENT FLUX: IMPLICATIONS TO REMOBILIZATION OF HG-CONTAMINATED SEDIMENT

In tide driven estuaries the tidal prism is an important process for the magnitude of the sediment flux. Since heavy metals adhere to fine sediment, the sediment path dictates the contaminant's fate. Alviso Slough in South San Francisco Bay is experiencing restoration of adjacent salt ponds into tidal ponds and salt marsh, which is having the effect of increasing the tidal prism and remobilizing mercury-contaminated sediment. We used a process-based model for investigating the impact of tidal prism increase on both sediment flux in the slough and the remobilization of contaminated sediment. Our results demonstrate that in Alviso Slough, the increase of tidal prism leads to erosion of the slough, increasing sediment export to the bay. Most of the exported sediment are due to erosion close to the mouth of Alviso Slough. Sedimentation in the slough decreases with increased tidal prism as a result of the changes in sediment flux and the import of sediment into the restored tidal ponds. Within a 3 month simulation, restoration of salt ponds to tidal exchange with the bay remobilizes about 3 kg of mercury. Most of the sediment remobilized from upstream, and mid-slough sediment is trapped inside the ponds, which to a large extent minimizes the export of this higher contaminated sediment into the South Bay.

This chapter is based on:

Achete, F. M.; Reys, C.V., van der Wegen, M.; Roelvink, D., Foxgrover, A., Jaffe, B.: Impact of a sudden tidal prism increase in estuarine sediment flux: implications to remobilization of Hg-contaminated sediment. Estuaries and Coasts (submitted).

4.1   Introduction

Estuaries are constantly under pressure to accommodate human needs and tend to be heavily impacted by the effects of urban development. Marshes are common habitats in estuarine regions, acting as protective buffer zones for the coastal lands from oceanic impacts. Marshlands are often diked to create new dry land for agriculture, salt mining, urban and industrial land (Boesch et al., 2001; Jackson et al., 2001). As more than half of the world population lives within 100 km of the coastline, sea level rise has implications for coastal safety. In this context, marsh restoration projects become an option to mitigate possible impacts of sea level rise.

Marshland restoration projects have been conducted worldwide e.g. in Washington (Yang and Wang, 2012), New York (Montalto and Steenhuis, 2004), Le Havre (Scherrer, 2006) among others. Since the 1970s, San Francisco Bay has experienced 45 tidal marsh restoration projects (Williams and Faber, 2004). South San Francisco Bay (South Bay) is part of the largest tidal wetland restoration project on the west coast of United States (http://southbayrestoration.org/). The project aims to allow tidal penetration inside the former industrial salt ponds to restore 60 km² of tidal wetlands (MacBroom, 2000).

Heavy metal, such as mercury (Hg), adheres to fine sediment. Despite its toxicity and bio-accumulation, Hg is observed in many estuaries (Feng et al., 2014), including San Francisco Bay. New Almaden, once the largest historical Hg mining area in North America, drains into South Bay (Ackerman and Eagles-Smith, 2010). After Alviso complex salt ponds were hydrologically connected to the South Bay and sloughs, upward trends of Hg were found in biota(Ackerman et al., 2013; Marvin-DiPasquale and cox, 2007b). This trend indicates that the salt pond opening induces bed scouring; therefore buried contaminated sediment is remobilized, and biota becomes exposed to it.

Restoration projects consist of breaching dikes and allowing the currents to bring sediment into the ponds in order to rebuild marshland. Similar to port development, and channel deepening, such projects, abruptly change estuarine geometry and tidal prism, modifying the hydrodynamic forcing and consequently the fine sediment flux (Williams and Faber, 2004). Equilibrium between the inlet cross-section area (A) and tidal prism (P) has long been studied and translated in the formula:

$$A=CP^q \hspace{8cm} \text{(Eq. 4-1)}$$

Where C and q are empirical parameters dependent on the estuary (Jarrett, 1976; Powell et al., 2006; Savenije, 2015; Stive et al., 2011; Van de Kreeke and Robaczewska, 1993); it follows that the increase in tidal prism leads to erosion at the mouth. The relationship for equilibrium velocity at the mouth (u_e) is:

$$u_e=pi^*P/T^*A \hspace{7cm} \text{(Eq. 4-2)}$$

Where T is tidal period. Replacing A in the first equation and considering q < 1 (Stive et al., 2011), shows that u_e increases with increase of tidal prism. Based on analytical solutions and data analysis, Friedrichs (2010) and Winterwerp et al. (2013) discussed the positive feedback between channel deepening and increase of SSC due to large modifications in tidal asymmetry which shift dynamics to a hyper-saturation environment. Little is known about the impact of these openings on the entire system's sediment dynamics including changes in flux and erosion/deposition area.

The aim of this work is to investigate the change in sediment dynamics in a tidal channel by an abrupt increase of the tidal prism, in this case as a result of tidal wetlands restoration. The work is done by applying a 3 and 2-dimensional horizontal process-based, numerical model (Delft3D FM, Kernkamp, 2010). The 3D model solves the complete shallow water equations while the averaged in the vertical (2DH) model solves the 2D vertically integrated shallow water equations, both coupled with advective-diffusive transport. We selected the Alviso Slough complex, South San Francisco Bay, California, as a study case since a major marsh restoration project is underway in the area, it contains contaminated sediments from legacy mercury mining, and observations of the study area pre- and post-opening are freely available. We analyze a) the impact on tidal propagation, b) changes in sediment flux along the channel, c) transport of mercury-contaminated sediment and d) estimate the volume of mercury deposited in restoration areas and in the Bay.

The San Francisco Bay is covered by a large survey network with freely available data on river stage, discharge and suspended sediment concentration (SSC) and other parameters from the USGS (nwis.waterdata.usgs.gov), Californian Department of Water Resources (http://cdec.water.ca.gov/) and National Oceanic and Atmospheric Administration (http://tidesandcurrents.noaa.gov/). The continuous SSC measurement stations are periodically calibrated using water collected in situ; that is filtered and weighed in the laboratory. In addition, the Bay-Delta system has high resolution (10m) bathymetry available for all the channels and bays (http://www.D-Flow-baydelta.org/).

4.2 Study area

Alviso Slough complex is located in the southernmost range of San Francisco Bay (SFBay) (Foxgrover, 2007), which is the largest estuary on the U.S. west coast (Fig 4-1). The system has a complex geometry consisting of interconnected sub-embayments, channels, rivers, intertidal flats, and marshes. It is surrounded by heavily urbanized area including San Francisco and the Silicon Valley. In this work, we focus on the ponds A5, A6, A7 and A8 that are adjacent to the Alviso slough (Fig 4-1). The levee between Pond A6 and Alviso slough was breached in two locations in 2010. The remaining ponds have water flow control structures, or openings, which allow managers to control the degree of exchange between the ponds, the Bay, and Alviso slough; the degree to which these control structures are open varies by year as described by Shellenbarger (2015). Ponds A5, A7, and A8 are interconnected by internal levee breaches.

In San Francisco Bay, water temperature and salinity have seasonal variability. During summer, the water temperatures are higher (~20 °C) while salinity increases to ~30 psu at Coyote Creek. During winter time, water temperatures drop to ~10 °C, at the same time the wet season starts, so salinity is lower (~10 psu). The tributaries typically have low discharge with winter peaks discharge lasting a couple of days (1-4 days). The main freshwater tributaries discharging in the area are Guadalupe River, which ends at Alviso Slough and has summer discharges of 0.85 m³s⁻¹ and winter peaks of 90 m³s⁻¹; Coyote Creek has summer discharges of 0.5 m³s⁻¹ and winter peaks of 20 m³s⁻¹ (Fig 4-2). Artesian Slough (2.6 m³s⁻¹ average discharge) and Moffett channel (1.3 m³s⁻¹ average discharges) receive water from municipal wastewater treatment plants (http://cdec.water.ca.gov/).

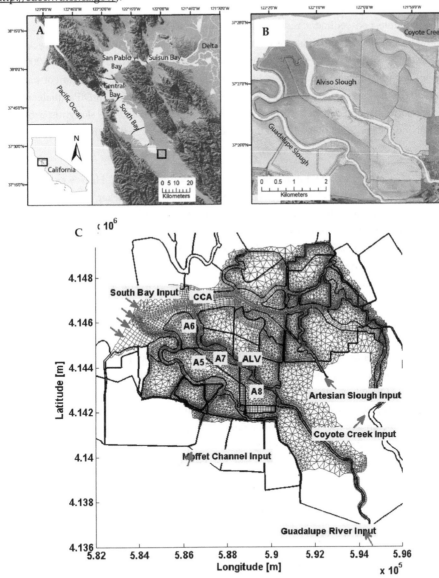

Figure 4-1: A) Alviso Slough complex study area location map.B) Zoom to the Alviso Slough, black rectangle at A. C) Model grid in blue, black lines delimitate the levees, red squares denote calibration stations, and the arrows the boundary conditions. Ponds A5, A6, A7, and A8 are indicated by the names and boundaries by red arrows.

Tidal energy propagates from the Golden Gate into South Bay up to Alviso slough. At Alviso the tide is mixed diurnal and semidiurnal, ranging from about 1 m maximum height during the weakest neap tides to 3.6 m during the strongest spring tides (Jaffe and Foxgrover, 2006). Despite receiving salt water from South Bay, most of the time Alviso is well mixed due to the low freshwater discharge from tributaries. However, during high peak flows it can sustain vertical stratification for the period of a day to a couple of days.

Local and remote bed sediment resuspension is the main contributor to suspended sediment concentration (SSC) levels in Alviso, as a result SSC is highly dependent on the tidal cycle. Mean Alviso Slough SSC is 180 mg L^{-1} varying from 50 mg L^{-1} at slack neap tide to peaks of 1200 mg L^{-1} during spring tide. Cohesiveness threshold is defined by 5% of clay in the sediment matrix. In Alviso Slough, the clay content is 26% and inside the ponds 33% (Marvin-DiPasquale and Cox 2007). Therefore, the top 2 meters of bed sediment in the ponds and the slough is constituted mainly of fine cohesive sediment. Hg adheres to fine sediment, making it possible to track it using sediment transport models.

Previous studies model sediment dynamics and scour in South Bay (H. T. Harvey & Associates et al.; McDonald and Cheng, 1996; Moffatt & Nichol Engineers, 2005; Bricker, 2003; Inagaki, 2000), but none focused on the tributaries or this restoration site. Data based studies have assessed the sediment budget at the restoration site (Callaway et al., 2013; Shellenbarger et al., 2004), and bathymetry analysis to define erosion and deposition patterns in the Slough (Foxgrover et al., 2011). Although measurements are discrete in space and/or time and do not explain the full sediment dynamics and deposition/erosion patterns, they are powerful tools to improve and give credibility to numerical models.

4.2.1 Model description

The numerical model applied in this work is Delft3D Flexible Mesh (Delft3D FM). DELFT3D FM allows straightforward coupling of its hydrodynamic modules with a water quality model, Delft-WAQ (DELWAQ), which gives flexibility to couple with a habitat (ecological) model. The coupling is further explained by (Achete et al., 2015). Deft3D FM is a process-based unstructured grid model developed by Deltares (Deltares, 2014). It is a package for hydro- and morphodynamic simulation based on a finite volume approach solving shallow-water equations by applying a Gaussian solver. The grid can be defined in terms of triangles, (curvilinear) quadrilaterals, pentagons, and hexagons, or any combination of these shapes. Orthogonal quadrilaterals are the most computationally efficient cells and are used whenever the geometry allows. Kernkamp (2010) and the DELFT3D FM manual (Deltares, 2014) describe in detail the grid aspects and numerical solvers.

The channels are defined by consecutive curvilinear grids (quadrilateral) of different resolution; each channel has at least 4 cells in the cross-channel direction. Channel junctions and ponds are defined by triangles (Fig 4-1). The average cell size ranges from 15x30m in channel area, to 120x120m in the ponds.

The base bathymetry and topography derives from interferometric side scan swath bathymetry collected in 2010 (Foxgrover et al., 2011), augmented by lidar of dry ponds and levees collected from June to November, 2010 (http://lidar.cr.usgs.gov/; Foxgrover et al, 2011) and single-beam echo soundings of submerged ponds collected in 2006 (Takekawa et al., 2010).The result was a detailed high-resolution digital elevation model (DEM) of the study area. In addition to the pre-restoration baseline surveys, an additional survey in 2012 documents erosion and deposition in Alviso Slough and intertidal flats in the Bay that can be used to verify model predictions (Takekawa et al., 2010).

Given a network of water levels and flow velocities (varying over time) DELWAQ can solve the advection–diffusion–reaction equation for a wide range of substances including fine sediment, the focus of this study. DELWAQ solves sediment source and sink terms by applying the Krone–Parteniades formulation for cohesive sediment transport (Krone, 1962; Ariathurai and Arulanandan, 1978) (Eq.4-1, Eq.4-2).

$$D = w_s * c * (1 - \tau_b/\tau_d), \text{ which is approximated as } D = w_s * c \qquad (4\text{-}3)$$

$$E = M * (\tau_b/\tau_e - 1) \qquad for \ \tau_b > \tau_e \qquad (4\text{-}4)$$

where D is the deposition flux of suspended matter (mg m^{-2}s^{-1}), w_s is the settling velocity of suspended matter (m s^{-1}), c is the concentration of suspended matter near the bed (mg m-3), τ_b is bottom shear stress (Pa), and τ_d is the critical shear stress for deposition (Pa), The approximation is made assuming, like Winterwerp (2006), that deposition takes place regardless of the prevailing bed shear stress. τ_d is thus considered much larger than τ_b and the second term in parentheses of Equation 3 is small and can be neglected. E is the erosion rate (mg m^{-2}s^{-1}), M is the first order erosion rate (mg m^{-2}s^{-1}), and τ_e is the critical shear stress for erosion (Pa).

4.2.2 Initial and boundary conditions

The model is initialized with the slack water at mean sea level, 10 psu for salinity, 10o C for temperature, SSC is 50 mg L^{-1} and at the bottom 2 m of fine sediment is available. To track the remobilization of mercury-contaminated sediment we define an initial distribution map considering five different concentration of THg (see section 4.3.8).

This model has five open boundaries, one bayward and four landward. Boundaries conditions are based on observations (Fig 4-1). At the seaward boundary we imposed hourly water level data from the Coyote Creek station (CCS, http://tidesandcurrents.noaa.gov/stations.html) (Fig 4-2); quarter-hourly salinity observation from the C17 station; quarter-hourly temperature and SSC observation from the Dumbarton Bridge station (DMB) (http://waterdata.usgs.gov/ca/nwis/) (Fig 4-2). The four landward boundaries are at tributaries. At Alviso Slough (GUA) and Coyote Creek (CCS), we impose quarter-hourly discharge and daily SSC and temperature observations (http://waterdata.usgs.gov/nwis) (Fig 4-2). Moffet channel and Artesian slough there is a constant discharge of 2.6 m^{3}s^{-1} and 1.3 m^{3}s^{-1}, respectively, from the wastewater plants.

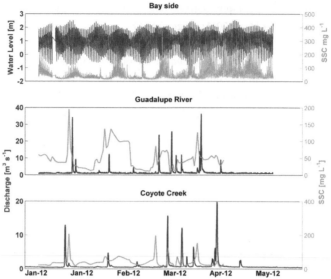

Figure 4-2: Boundaries Condition input for water discharge in blue and suspended sediment concentration in green.

The calibration and validation periods were chosen based on operations at salt pond A8 and data availability at the Alviso Slough station (ALV) (Fig 4-1). Observations at ALV started on March 2012 for water depths, discharge, velocities and SSC (Shellenbarger et al., 2015). The calibration period is from March to May of 2012. The validation period is from August to December 2012, which comprises high and low river discharges during spring and neap tide. Gates to Pond A8 were closed during both the calibration and validation periods. First, we ran the hydrodynamic model for the above mentioned periods to calculate water level, velocities, cell volume and shear stresses. Then, the hydrodynamic results were imported to DELWAQ to calculate SSC levels (Achete et al., 2015).

The discharges, water level, and SSC model results are compared to in situ measured data. The calibration process for the hydrodynamics includes a spatial variation of the friction coefficient and adjustments to bed level to match observed phasing of water levels and velocity. For the sediment model, the analyzed parameters are fall velocity (w_s), critical shear stress (τ_{cr}) and erosion coefficient (M).

The scenarios are defined in order to determine changes in sediment fluxes and erosion/deposition of mercury-contaminated sediment considering different gate opening scenarios and the influence of bathymetry changes.

4.3   Results and discussion

4.3.1 Hydrodynamic model

To study the impact of pond gate opening on the sediment dynamics, we first calibrated the hydrodynamic model at ALV station in terms of water level, discharge, and velocity.

Hydrodynamic calibration was carried out by systematically varying the value of the Manning's coefficient (man) from 0.012 s m$^{-1/3}$ to 0.035 s m$^{-1/3}$ (Fig 4-3). Statistical coefficients are useful tools to assess the best parameter settings. The bias (Eq. 4-5) indicates if the model systematically under- or overestimates observed SSC levels. The unbiased root-mean-square error (uRMSE') (Eq. 4-6) values the modeled variation compared to observations and the correlation coefficient (R) (Eq. 4-7). The observed mean discharge was -2.5 m^{3}s^{-1} (bayward) with a standard deviation of 19. The choice of the standard run (0.026 s m$^{-1/3}$) is based on statistical coefficients like uRMSE, bias and R (Table 4-1).

$$\text{Bias} = \bar{m} - \bar{r} \tag{4-5}$$

$$uRMSE' = \left(\frac{1}{N}\sum_{n=1}^{N}[(m_n - \bar{m}) - (r_n - \bar{r})]^2\right)^{0.5} \tag{4-6}$$

$$R = \frac{\left(\frac{1}{N}\sum_{n=1}^{N}[(m_n - \bar{m})(r_n - \bar{r})]^2\right)}{\sigma_m \sigma_r} \tag{4-7}$$

Where 'm' refers to the model result and 'r' to observations.

Table 4-1: Calculated uRMSE and bias for hydrodynamic calibration, comparison of different manning coefficients.

	uRMSE (m³s⁻¹)	Bias (m³s⁻¹)	R
Man = 0.014	59	14.2	0.26
Man = 0.026 (best)	10	-1.3	0.85
Man = 0.032	11	0.06	0.80
Man = 0.014 - 0.035 (tidal channel - flats)	30	-6.1	0.69

At ALV station, the tide propagates as a classical standing wave where the velocity leads water level by 90°, resulting in the highest ebb and flood velocities at MSL. The tidal asymmetry parameter (γ) proposed by Friedrichs and Aubrey (1988) and simplified by Friedrichs (2010) Eq. 4-8 is positive indicating flood-dominant estuary.

$$\gamma = \frac{a}{h} - \frac{1}{2}\frac{\Delta b}{b} = \frac{3.5}{2.5} - \frac{1}{2}\frac{40}{50} \tag{4-8}$$

Where 'a' is tidal elevation, 'h' is average channel depth, Δb is amplitude of the tidal variation in estuarine width and b estuary width, the over bar indicate average values. Alviso slough channel at ALV is constricted by high tidal flats and so $\bar{b}$ is 50 m and Δb is 40 m, considering 'a' 3.5 and 'h' 2.5 the resulting γ is 1.

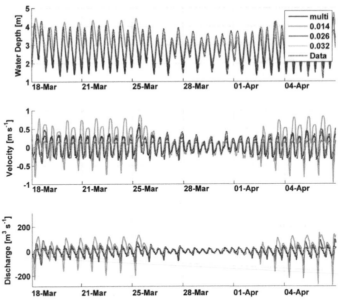

Figure 4-3: Hydrodynamics calibration at ALVS a) water level, b) velocity and c) discharges. The black dashed line denotes observed water depth, velocity and discharge and the colored lines denote calibration runs applying Manning coefficient 0.014 s m$^{-1/3}$ in the channel and 0.035 s m$^{-1/3}$ in the flats (multi), 0.014 s m$^{-1/3}$, 0.026 s m$^{-1/3}$ (best run) and 0.032 s m$^{-1/3}$.

As Alviso complex connects to South Bay and receives salt water, the first runs were performed in 3D with 10 vertical layers to define the system stratification. Richardson number ($Ri = \frac{\frac{g}{\rho}\frac{d\rho}{dz}}{\left(\frac{du}{dz}\right)^2}$) measures the stratification stability, if $Ri < 0$ the system has an unstable stratification and the water column will mix by itself, if $0 < Ri < 0.03$ stratification is not strong enough to damp the turbulence and the system is mixed, but if $Ri > 0.3$ the system is stratified and therefore needs a 3D model to simulate the dynamics (Dyer, 1986).

During peak events stratification can reach a Ri of 20; however these peaks do not last longer than 1 to 3 days. Disregarding river peak events, at ALV, the mean Ri is 0.13, lower than the 0.3 that is the limit to attenuates mixing allowing a 2DH modeling approach (Fig 4-4). Most of the time, we observe a salinity front propagation creating a well-mixed water column. In neap tide during short periods the salinity front propagates as salt wedge reflected by the small spikes in the Richardson number (Fig 4-4), Shellenbarger (2015) discuss further the impact of this stratification.

4.3.1 Sediment Calibration

The Krone-Partheniades equation leaves us with three calibration parameters. Considering a range of variation for each single parameter, sensitivity analysis requires a large amount of runs. Target diagrams summarize and combine the normalized by the standard deviation uRMSE* in the X-axes and the normalized by the standard deviation Bias* in the Y-axes, facilitating the best run choice (Jolliff et al., 2009) (Fig 4-5). By definition uRMSE* is positive, to determine if the model standard deviation is larger (X>0) or smaller (X<0) than the observation, the uRMSE* is multiplied by the standard deviation difference $\sigma_d = sign(\sigma_m - \sigma_r)$. The coefficients are normalized by the reference field standard deviation.

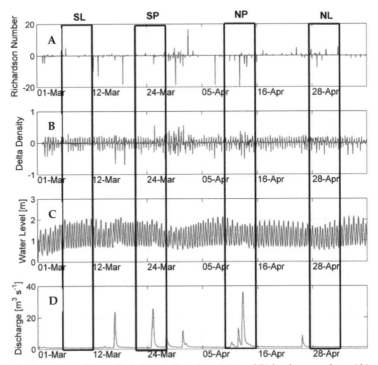

Figure 4-4: Hydrodynamic model result at ALV station a) variation of Richardson number within a month, b) shows the density difference between the top and bottom layer, c) the water level and the d) is the water discharge at GUA station, to highlight the peaks. Black rectangles indicating the hydrodynamic periods: spring tide low river discharge (SL), spring tide peak river discharge (SP), neap tide peak river discharge (NP) and neap tide low river discharge (NL).

Target diagrams have been mostly applied in biology and biochemistry (Edwards et al., 2012; Fraysse et al., 2013) and hydrodynamics (MacWilliams et al., 2015) studies. In these diagrams, a model that exactly reproduces the data is placed in the center, so the best performance run is located closest to the origin. The circle indicating the isoline of 1 in a normalized diagram delimitates where the difference in the mean value of model and observation is as large as the standard deviation of the observations. The goodness of the models assessed by a target diagram depends on the analyzed parameter. Studies analyzing water level, tidal velocity, salinity, and chlorophyll a have results placed inside the 0.5 isoline, while phytoplankton absorption is around the 1 isoline and nutrients around the 2 isoline, so far one study has applied target diagrams to SSC (Los and Blaas, 2010).

We applied a target diagram analysis to SSC with two objectives, first to define the best parameter settings and second to define the best output time-step. The initial parameter settings were $w_s = 0.25$ mm s^{-1}, $\tau_{cr} = 0.25$ Pa and $M = 10^{-4}$ kg m^{-2} s^{-1} (Manning and Schoellhamer , 2013). To determine the best parameter settings we compared 30 runs (Fig 4-5) from sensitivity analyses of each parameter with output every 15 min. Each parameter is varied separately, ranging w_s by 0.1-1.0 mm s⁻¹, τ_{cr} by 0.1-0.5 Pa and M by 10⁻⁶-10⁻³ kg m⁻² s⁻¹. Runs with multiple parameter changes were tested in the target diagram to determine which setting was closest to the origin, resulting in the following best run (standard run) with settings w_s= 0.5 mm s⁻¹, τ_{cr} = 0.25 Pa and M = 2x10⁻⁵ kg m⁻²s⁻¹ (Fig 4-5 and Fig 4-6).

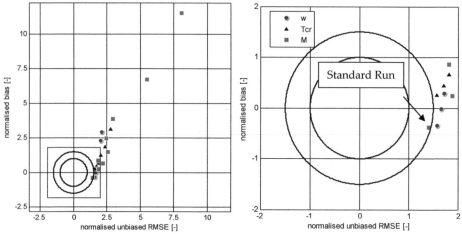

Figure 4-5: Target diagrams for suspended sediment concentrations at Alviso Station (ALV) located at mid-slough. The calibration runs, with 15 minutes output, are shown. The symbols indicate the runs; blue circles indicate variation in w_s, black triangles are variation in τ_{cr} and the red squares are variation in M. The circle indicating the isoline of 1 in a normalized diagram delimitates where the difference in mean value of model and observation is as large as the standard deviation of the observations.

The large scatter in the target diagram shows the sensitivity of the model to changes in parameters (Fig 4-5). M, represented by red squares, is the most sensitive parameter with a bias of 12 mg l⁻¹ and uRMSE of 8 mg l⁻¹. The spread is close to a linear progression where bias is more affected than uRMSE, thus, an increase in M shifts the SSC curve upwards and increases the variability. The high sensitivity to M is caused by tidal resuspension in the slough. Critical shear stress has a similar behavior to M but is less sensitive. Decreasing w_s increases bias but does not significantly affect uRMSE.

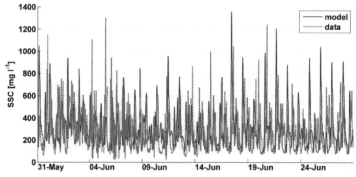

Figure 4-6: SSC calibration curve, standard run. In blue the model results and in dashed red observations.

Having high-frequency observations (every 15 minutes) allows assessment of lower model output frequencies; if lower frequencies can accurately describe the system dynamics, they are preferred as they are easier to work with. We tested the calibration with model outputs of 15, 30, 60 and 120 minutes. We also tested filtering the model and observations only to allow frequencies lower than 12, 24 and 48 hours.

The system we analyzed has major tidal variations. As a result, output frequencies up to 2 hours have similar performance, while smaller time steps do not improve results. Filtering the results consists of shaping the signal spectrum. In this case, we use a Butterworth low pass filter, where variation in the signal higher than 12h, 24 h and 48 h were disregarded when comparing model and data. The filtering removes variability related to the daily tidal cycle maintaining the neap and spring variation. The SSC difference from spring to neap in the data is larger than in the model results, for this reason, the model uRMSE* in the filtered results are negative (uRMSE standard deviation difference signal). Besides underestimating the results (increased bias), the performance of the low-resolution analyzes are comparable to the parameter variations.

4.3.2 Sediment dynamics

Shellenbarger et al. (2015) analyzed sediment flux between ALVS and GUA based on observations. The present model has GUA station as a boundary condition for water discharge and SSC and ALVS as calibration stations where modeled water discharges and SSC levels agree with observations. This section explores to what extent Shellenbarger et al.'s (2015) sediment fluxes are similar to ones from our modeling effort and extends the analysis to the entire slough. For the sake of clarity, positive discharge and flux are flood directed indicating a landward import of sediment; negative discharge and sediment flux are ebb directed indicating bayward export of sediment bayward

To describe the channel discharge, we analyze 5 cross-sections (Fig 4-7). During this period, pond A6 gates are open. The first is GUA at the landward boundary, the second is at ALVS station after A8 notch, the third ALVSOUT is at middle slough, the fourth between A6 pond breaches L10, and the fifth is at the mouth (MOU). The slough attenuates and reflects the tidal wave so that no tidal variation is observed at GUA.

For the whole slough the cross-section tide residual discharge varies from -1.4 m^3s^{-1} at GUA up to -2.1 m^3s^{-1} at MOU. However, at all locations, except GUA, the discharge of the tidal wave dwarfs the residual discharge. At GUA station, the water discharge is unidirectional bayward, at ALVS, the tidal wave starts influencing the discharge (σ = 3.4 m^3s^{-1}). At ALVSOUT, tidal influence is about 10 times larger than the river discharge (σ = 16 m^3s^{-1}) and at the MOU σ is 70 m^3s^{-1}.

This difference in tidal strength combined with peak river discharge defines the sediment transport in the slough. We analyze cumulative sediment flux over a week for each segment of the slough under 4 hydrodynamic conditions: neap tide and low river discharge (NL), neap tide and peak river discharge (NP); spring tide and low river discharge (SL); and spring tide and peak river discharge (SP) (Fig 4-4). We also calculate the cumulative sediment transport over a 3-

month period encompassing all the aforementioned hydrodynamic conditions to define the net sediment transport.

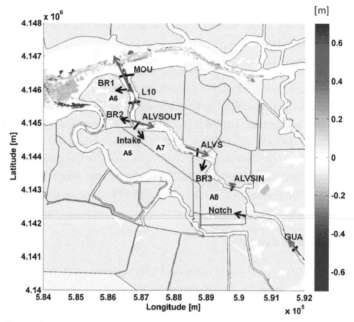

Figure 4-7: Difference between 2010 and 2012 measured bathymetry. Red indicates deposition. The black lines indicate the model cross sections, the red arrows at the cross-section indicate the direction and magnitude of the sediment flux in the slough. The black arrows indicate the breach locations pointing to the positive flux direction.

Close to the landward boundary the GUA cross-section unidirectional flow implies bayward sediment flux for all the hydrodynamics conditions; we can observe that the blue line in Figure 4-8 (a, b) always exports sediment and has no tidal influence. At GUA the river dynamics modulates sediment transport, during NL and SL conditions the mean sediment flux is -7×10^{-4} kg s^{-1}, during 1 to 3 days of river peak events (NH and SH) the sediment flux reaches -22 kg s^{-1}, discharging ~180 t in less than 3 days (Fig 4-8). Peak events are observed as steps in the cumulative sediment transport (Fig 4-8a).

At ALVS bidirectional sediment transport reflects the tidal influence in the region (Fig 4-8b), and the mean sediment flux is 6×10^{-4} kg s^{-1} landward. Flood dominant tidal wave, with flood duration shorter than ebb duration (Friedrichs and Aubrey, 1988; Horrevoets et al., 2004), forces landward sediment flux where the 3-month cumulative transport adds up to 1,300 t, the same as calculated by Shellenbarger (2015). During peak events, the bayward transport induced by river flow overcomes the flood-tide induced landward transport. River peak discharge (SP and NP) is able to change the sediment flux direction and flush sediment bayward, behavior also observed by Shellenbarger et al. (2015) (Fig 4-8). The duration of the sediment transport direction change is more related to the peak intensity and subsequent tide cycle than the tide cycle during the peak event. At NL the net sediment flux is low; if a neap tide follows the peak event, the exporting trend in sediment flux is sustained during the neap cycle, but if the peak is followed by a spring cycle, the tide forces back the landward flux.

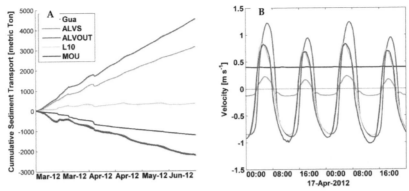

Figure 4-8: A) Cumulative sediment transport for each cross-section, positive transport is importing towards the slough and negative exporting towards the bay. B) Velocity for each cross-section at 17th of April, during spring tide and low river discharge.

At ALVSOUT, the flood-tide asymmetry has a major impact on sediment flux. At ALVSOUT, the mean flux is 7.5 x10^{-4} kg s^{-1} and for all the conditions the net sediment flux is landward (Fig 4-9). The highest rate of landward transport is during low river condition as SL and NL when the cumulative transport is ~300 t, because peak events are able to reduce the landward transport by half (Fig 4-9). Between GUA and ALVS, the forcing changes from exporting to importing leads to a sedimentation area, which in this case is imported into the ponds (Fig 4-9). This process of tidal forcing induced sedimentation is observed in Petaluma River (Ganju et al., 2004) and in the Delta (Achete et al., 2016).

Closer to the mouth but still before the last breach of pond A6, cross-section L10 is in the middle of another transition zone of importing to exporting of sediment. At this point, the tidal wave is not as distorted as further inside the slough (Fig 4-9), and it is possible to observe a sediment export trend. Notice that at L10 the net transport is one order of magnitude lower than elsewhere with cumulative of 300 t, and the standard deviation is 20 t. Considering an error bar of +/-20t, the cumulative net transport for the periods NL, SL, NP, and SP is statistically zero.

The sediment dynamics at the mouth (MOU) are dominated by the ebb dominant tidal wave. Here, ebb velocities are higher than flood, exporting sediment from the slough to the bay (Fig 4-8 a, b). The transport is modulated by the tides. Therefore, river peak events have little influence on the sediment flux. During neap tide, the net sediment transport is half of the transport during the spring tide and the total net transport exports 2,200 t of sediment bayward (Fig 4-9).

The most remarkable difference between peak and low river discharge is in ALVS cross-section, where the sediment flux changes from exporting during peak flows to importing during low flow. Considering the 3 months cumulative sediment transport for the 5 cross-sections it is possible to define erosion and deposition trends at the slough. The most upstream stretch from GUA up to ALVS has a deposition of 2,800 t, at the mouth erosion indicated by the model results is observed in the bathymetry soundings from 2010 to October 2012 (Fig 4-9). The ponds sediment import is of the same order of magnitude as slough transport, and in most of the cases surpass it, indicating an erosion in the slough. The pond sediment import is almost constant across hydrodynamic conditions, apart from the period of Spring tide Low discharge condition when Br01 and the intake import ~400 t each. For the simulation period between ALVS and L10, 9,000 t is eroded from the slough, becoming sediment trapped in the ponds.

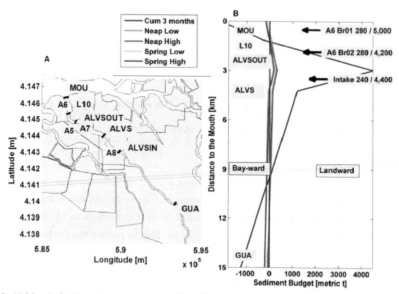

Figure 4-9: A) Map indicating the cross-sections, the red arrows at the cross-sections indicate the direction of positive flux and the black arrows indicate the breach locations pointing to the positive flux direction. B) Indicates the sediment flux for each hydrodynamics condition (colored lines) and cumulative for 3 months (blue line) when the A8 notch is closed. The arrows indicate an average 3-month cumulative sediment flux into the pond for the different scenarios; scenarios that deviate from these values are specified in the text.

The simulated period encompasses dry and wet periods, and we showed that the peak discharges considerably change sediment dynamics in the upper slough. To extrapolate the sediment budget for an entire year, one should only consider low river discharge during spring and neap tide.

4.3.3 Tidal prism step increase

In Alviso, the sudden increase of tidal prism is caused by pond breaching and in this section, we discuss different breaching/opening scenarios (Table 4-2). The boundary input is from February to March 2010 when pond A8 was closed. In this section 1 cross-section (ALVSIN) was added to explain better the influence of the breaches (Fig 4-9).

Tidal amplitude times the planform area gives the tidal prism. Tidal prism increase leads to higher velocities at the estuary mouth (Stive et al., 2011). Alviso slough and tidal flats have an area of 1.3 km², for a tidal amplitude of 2 m the tidal prism is 2.6 x 10⁶ m³. Pond A6 has an area of 2.4 km² and a tidal amplitude of 1.5 m resulting in a tidal prism of 3,600 m³. The A8 complex includes A5, A7 and A8 and has an area of 5.6 km². The tidal average amplitude in the A8 complex is of 1 m; thus, the tidal prism is of 5,600 m³. The opening of the 2 systems increases the initial tidal prism by 350%. Applying the updated Escoffier equation (Eq. 4-1 and Eq. 4-2), the estuarine cross-sectional area increases with the increase of tidal prism until reaches a new equilibrium. The mouth erosion has already been observed in situ (Fig 4-7) between MOU and Br02 where approximately 0.7 m was eroded over 2 years (Foxgrover et al., 2011).

The empirical equations help to explain the consequences at the mouth but not the changes in fluxes inside the estuary. The scenarios were designedto analyze the changes in velocity and sediment transport. In scenario 01 (sc01), representing pre-restoration conditions, all the ponds are closed (Table 4-2). In scenario 02 (sc02) pond A6, which includes Br1 and Br2, and the intake in A7 are open (same openings as the previous session). In scenario 03 (sc03) ponds A6, the intake at A7 and A8 notch is 5 m wide, scenario 4 (sc04) has the same openings as sc03, but the A8 notch is 15 m wide. For scenario 5 (sc05) a new hypothetical breach (Br03) at A8 pond is considered. For all the scenarios the initial bed composition is the same. We allow for 1 month of spin up time to adjust to the hydraulic conditions.

Table 4-2: Scenario description.

Scenario	Openings	Tidal Prism
1	All closed	2.6×10^6 m^3
2	A6 + intake	3.6×10^6 m^3
3	A6 + intake + A8 Notch - 5 m (Around 15 ft)	5.6×10^6 m^3
4	A6 + intake + A8 Notch - 15 m (Around 40 ft)	5.6×10^6 m^3
5	A6 + intake + A8 Notch - 15 m (Around 40 ft) and open breach at Pond A8	5.6×10^6 m^3

First we discuss the changes in the tidal propagation in the slough among the scenarios (Fig 4-10) and later the impact on sediment dynamics (Fig 4-11). At MOU, from sc02 to sc05 ebb velocities are the most impacted by the larger tidal prism increasing from 0.5m s^{-1} to 1.0m s^{-1}. The tidal wave becomes more distorted delaying the peak ebb velocity by 1 hour. The delay is reflected in the water levels and more asymmetric discharges (Fig 4-10).

At ALVSOUT, we observe skewed, smaller amplitude in water level. The smaller amplitude is due to water deviation to the ponds. The velocity delay in relation to water level observed at MOU for sc01 is accentuated for the scenarios, leading to asymmetry in discharges. At ALVSIN, the same behavior is observed as ALVSOUT, with smaller water level amplitude accentuated and in sc05 the low water is larger than the mean water level in sc01 (Fig 4-10). The friction in the channel attenuates the tidal wave, decreasing velocities and discharges.

During flood tide, part of the flux is deviated inside pond A6 through Br01 and Br02 generating small eddies inside the pond. The flux enters pond A7 and A8 through the A8 notch since the velocities landward are already smaller than at the mouth, the velocities inside these ponds are less than 0.2 m s^{-1}.

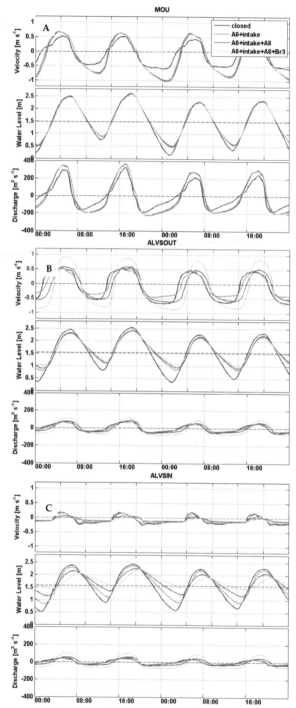

Figure 4-10: Cross-section averaged velocity, discharge and water level time series for sc 01 (dark blue), sc02 (green), sc03 (red) and sc05 (light blue). a) Seaward at MOU, b) ALVSOUT and c) landward ALVSIN cross-section.

Changes in hydrodynamics lead to changes in sediment flux. We observe a progressive change in sediment dynamics along the slough when more openings are included. In sc01, the exporting of sediment is observed from GUA cross-section until ALVSIN, when flood dominant tide starts importing sediment (1,300 t). In sc01 the deposition area extended from GUA to ALVSOUT, since there is no open breach the deposition area is between the dark blue line and the zero line (dashed black) (Fig 4-11). The area between ALVSOUT and MOU is erosive with a sediment deficit of 3,400 t because while ALVSOUT is importing sediment (1,200 t) MOU is exporting to Coyote creek 2,200 t.

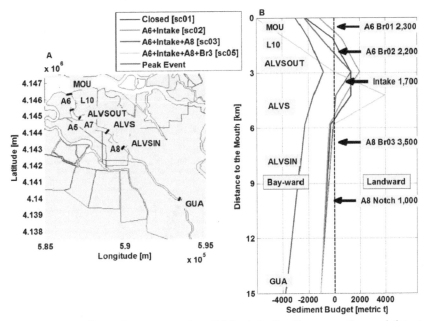

Figure 4-11: Scenarios sediment flux representation. A) Map indicating the cross-sections and the arrows the breaches. B) Each scenario is a different color. The arrows are the scenarios average sediment flux into the ponds, otherwise specified in the text.

As expected the pond opening does not influence the river input boundaries so at GUA sediment flux is always baywards and no remarkable changes are observed landward of the A8 notch. In sc02, the opening of the water intake at A8 and breaches at A6 decrease the deposition between ALVSIN and ALVS and the sedimentation zone decreases and shifts bayward (Fig 4-11). Most of the sediment entering the ponds are from local bed scour since the river sediment flux entering at GUA is 1070 t for all scenarios, at the mouth the net sediment flux is exporting, and the total pond intake is about 7000 t.

Comparing sc01 and sc03, between ALVS and ALVSOUT while in sc01 sediment flows from one station to the other while in sc03 the A7 intake import the sediment to pond A7 creating an erosion of 700 t. At L10, the first opening (sc02) reverts the net flux landward but with further openings (sc03 and sc05) it exports again (Fig 4-11). Comparing sc03 with sc04 shows that the size of the notch at A8 does not affect the tidal prism, so the changes in sediment flux are not significant. Still, the larger notch allows for further intrusion of the sediment deposition plume into pond A8 (Fig 4-12).

The sediment import into the pond (black arrows in Fig 4-11b) is not affected overall by the different scenarios. The pond A6 breach Br01 imports about 2,300 t and Br02 about 2,200 t. The import to the pond A7 intake depends on the A8 notch opening, when A8 notch is open ~1,750 t enters the intake and when the notch is closed approximately 2,500 t enters. The notch imports 1,000 t in sc03 and sc04 but with the new hypothetical breach (Br03) at A8, it imports 3,500 t, because it is located in the deposition area of the Slough.

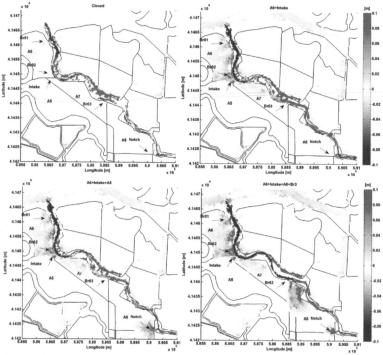

Figure 4-12: Erosion deposition maps for the scenario runs, sc01 (closed), sc02 (A6+Intake), sc03 (A6+Intake+A8), sc05 (A6+Intake+A8+Br3).

In sc05, we simulate the impact of a hypothetical breach at A8 (Br03) on the sediment flux. In sc05 the combination of longer ebb period, higher flood and ebb velocities and shorter slack water increase the import of sediment in ALVS as well as increase export at L10 and MOU (Fig 4-11). Br03, located in the middle of the slough deposition area, stimulates sediment import into pond A8. The new opening decreases the sediment import through the A8 notch and at the same time enhances erosion downstream of Br03 (Fig 4-12). The hypothetical breach shows the importance of analyzing breach locations individually, in the Br03 case the location is essential for sediment import to the ponds and slough erosion, since it is located in the slough deposition zone between ALVSIN and ALVS, where the sediment flux converges.

Figure 4-12 shows maps of erosion and deposition for the different scenarios. Pond opening enhances erosion at areas close to the breaches, downstream from the breach and at the slough mouth (Fig 4-12). The rest of the slough presents sedimentation. This pattern agrees with the observed bathymetric change between 2010 and 2012 (Fig 4-7). The ponds act as sediment sink (Fig 4-11 and Fig 4-12), once the ponds are opened, the slough deposition area starts to erode due to sediment being trapped inside the ponds and higher velocities due to the larger tidal prism.

4.3.4 High river discharges events implications

The 3-month simulation consisted of dry period and average peak discharges during neap and spring tide. However, Alviso Slough also experiences extreme high river discharge events, especially during El Niño years. The extreme events last about a week and the peak water discharge reaches 150 m^3s^{-1}. We set a different simulation for a high peak event included the same pond openings as sc03, breaches Br01 and Br02 at pond A6, the notch at pond A8, and the intake.

Larger river discharge events shift the net slough sediment flux to fully exporting towards the bay for all the cross-sections. Despite the exporting trend, the slough keeps the same deposition/erosion pattern previously described, with sediment flux convergence conversion near ALVOUT. During normal conditions, ALVOUT imports sediment at an average rate of 200 t per week. During the peak event week, almost 8 times more sediment (1500t) is exported. After the peak event, the slough returns to the import/export pattern previously observed but to have the cumulative transport return to predominantly importing, it would require about 2 to 3 months without a major river event.

4.3.5 Sensitivity in sediment flux

In session 4.3.2, we presented the best run based on RMSE and bias. Here we discuss the impact on cumulative sediment flux in the worst case for the most sensitive parameter the erosion coefficient (M). In the standard run M is $2x10^{-5}$ kg $m^{-2}s^{-1}$, in the target diagram (Fig 4-5) the least correlated series correspond to $M = 10^{-3}$ kg $m^{-2}s^{-1}$.

Overall the higher M leads to larger sediment flux in the slough because of the higher SSC, and the magnitude of the difference depends on the region. Close to the upstream boundary, the sediment transport is 20% smaller in the standard run than the higher M scenario. The region where the tidal currents start to modify the river discharge, between ALVSIN and ALVS, is the most impacted. The higher SSC in the test case (higher M) facilitates the export of sediment even during low river discharge, exporting more sediment than in the standard case. In the standard case there is a deposition area where the tidal discharge decreases river discharge the sediment transport is almost zero. In the test case, SSC is higher and carrier further, the deposition zone is shifted bayward. From ALVOUT until the slough mouth, the standard and the test case have the same sediment flux direction with the difference being the test case the SSD is the double that of the standard case.

4.3.6 Morphological updating

Changes in tidal prism affect the tidal propagation inside the slough, changing sediment flux and inducing erosion and deposition. In the model erosion is most intense at the breaches and at

Alviso's mouth, with an average rate of ~0.02 m per month (0.07 m for the 3-month simulated). The model output volume is similar to the volume calculated from observed changes in bathymetry from 2010 and October 2012. The high rate of morphological change leads to further changes in sediment flux since the bathymetry dictates the velocities inside the slough and, therefore, the sediment flux.

Following Escoffier (Elias and van der Spek, 2006; Stive et al., 2011; van de Kreeke, 1985) (Eq. 4-1 and 2), if no other pond is open, the inlet cross-sectional area tend to convert to an equilibrium, decreasing erosion rates, thus decreasing sediment flux in the slough which is closely related to bed resuspension. The present model does not consider morphological changes, thus, the new equilibrium will not be reached, and high sediment fluxes are persistent in the model. Not considering morphological changes, makes this model unsuitable for analyzes longer than 5 years. It is expected that the rate of mercury remobilization and erosion inside the ponds will decrease as the slough approaches a new equilibrium.

4.3.7 Tracking mercury-contaminated sediment

Under the assumption that mercury is and remains attached to (transported) sediment, tracking sediment particles will give an indication of the re-allocation of mercury contamination. In this section, we discuss the impact of the different scenarios on the redistribution of mercury-contaminated sediment. Deep cores of bed sediment show a spatial variability of THg concentration. We defined initial mercury concentrations based on Fregoso et al. (2014) that vertically and horizontally interpolated concentrations from existing sediment core data. Three slough stretches are defined based on THg concentration as high (757 ng/cc) between cross-sections ALVSOUT and ALVS, medium (370 ng/cc) between cross-sections ALVS and GUA and low (168 ng/cc) between cross-section ALVSOUT and MOU. These values are the result of a vertical profile average over 2 m. In the model, the concentrations are defined as fractions fr03 (medium), fr04 (high) and fr05 (low) (Fig 4-13). These fractions can be tracked during the simulations. Inside the ponds A6, A7 and A8 no initial sediment is available, all the other areas the bay sediment fraction 2 (fr02) is defined as initial condition, fraction 1 (fr01) is the Guadalupe River input and is not defined as bed initial condition.

The river load is very low compared to bottom remobilization and it mainly deposits landward of the A8 notch. The bay sediment is mostly carried to Coyote Creek or inside pond A6 when open. Transport towards the A6 pond almost doubles at sc03, sc04 and sc05 because of the larger prism and higher velocities.

Fr03 is set on the most upstream portion of the slough (Fig 4-13). In sc01, 1,300 t of fr03 (or 0.35 kg of THg) is transported all the way to Alviso's mouth towards the Bay. However, with the pond open (sc02-05) the majority of sediment remains trapped at the initial condition position or is transported inside pond A8 regardless the notch size (~ 1,000 t, 0.26 kg). Sc03 is the only scenario where fr03 is found in pond A6 (Fig 4-13). For all the scenarios less than 0.1% of the sediment exported to the bay consists of fraction 3.

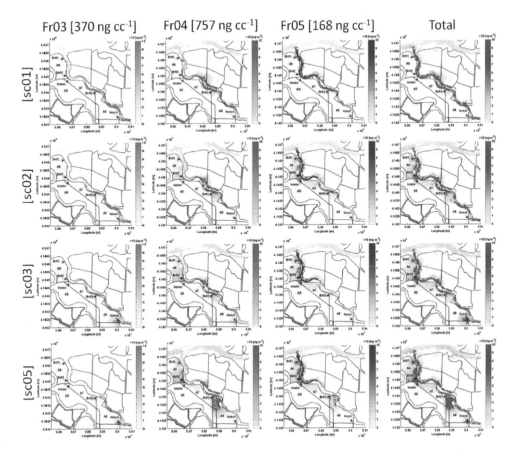

Figure 4-13: Concentration maps of mercury-contaminated sediment for fraction 3 (medium - green), fraction 4 (high - red). fraction 5 (low - blue), and the combination of all fractions in the last time step of the simulations sc01 (closed), sc02 (A6+Intake), sc03 (A6+Intake+A8), sc05 (A6+Intake+A8+Br3).

The middle slough fr04 has similar behavior as fr03 remaining mostly in the slough (Fig 4-13). On sc01, fr04 is more easily exported bayward (900 t, 0.52 kg) than in the other scenarios. On sc03 the convergence point shift transports 2,200 t (1.28 kg) of fr04 inside A6. In sc05 the new breach is in front of the previous deposition area transporting 3,500 t (2.03 kg) of fr04 inside A7 and an additional 4,700 t (2.73 kg) to A6.

Fr05 composes the slough bed sediment closest to the mouth and, therefore, most stirred and transported by tides (Fig 4-13). For all the scenarios fr05 deposits upstream in Alviso slough, inside the ponds and downstream in Coyote Creek. The larger tidal velocities triggered by the larger tidal prism on sc05 transports fr05 even further bayward and upriver of Coyote Creek. Almost 60% of the sedimentation observed inside ponds A6 and A8 consist of fr04, about 4,000 t (2.32 kg) for each pond. Fractions 04 together with fr02 constitute 80% of the transported sediment through the SEA cross-section at the bay boundary.

4.4 Conclusions

Our results show that increased tidal prism modifies the sediment flux in the estuary. The abrupt change in tidal prism caused by pond opening leads to higher tidal velocities inside the estuary and to an overall increase of erosion, but this change is not uniform along the slough. It increases erosion at the Slough's mouth and decreases the deposition area in the mid-slough, even leading to erosion downstream from the breaches.

The mercury is modeled as a tracer attached to the fine sediment. During normal hydrodynamics conditions, tidal current dominates sediment transport. This modeling shows that opening ponds in the Alviso complex lead to deposition inside the ponds of about 9,000 t of sediment in a 3-month simulation. Overall the pond openings tend to keep the most contaminated sediment, the upper and middle slough sediment, inside Alviso complex as a result of the deposition in the slough and ponds. In contrast, sediment close to the mouth is widely spread throughout the slough and Coyote Creek post pond openings. In the scenario with the most openings, about 3 kg of THg was deposited inside the ponds over the 3-month simulation.

Projects like expansion or deepening of ports result in an increase in a tidal prism that should be carefully studied beforehand to avoid erosion of sensitive areas and remobilization of contaminated sediment.

To allow a better understanding of the system and enable further modeling efforts, it is important to maintain the continuous measuring station in the middle of the slough, and an additional station at the mouth would be beneficial to verify the model results. As the bathymetry observations show, there is a significant change in bathymetry due to scouring at the pond breaches. A morphological model that includes feedback between morphological change and hydrodynamics will increase the understanding of the system and may be able to predict whether a new equilibrium will be reached and how long it would take to reach such equilibrium.

Despite tides defining sediment flux in the slough, high river discharge events can briefly overcome the dominance of tides shifting the estuary from an importing to an exporting system. To predict the long-term flux, morphologic change, and remobilization of contaminated sediments in tidally dominated small estuaries, both event and non-event periods must be accurately modeled.

Acknowledgments

The research is part of a multidisciplinary study of the restoration of South San Francisco Bay salt ponds. The authors acknowledge the California Coastal Conservancy and the US Geological Survey Coastal and Marine Geology Program for making this research financially possible. Laura Valoppi, Marc Marvin De-Pasqual and Greg Shellenbarger supplied invaluable input concerning the hydrodynamics and mercury dynamics of the area. Rusty Holleman and Tara Schraga provided helpful reviews of the manuscript. This research was also partially funded by the Brazilian Government via Capes agency.

5

How important are climate change and foreseen engineering measures on the sediment dynamics in the San Francisco Bay-Delta system?

Many estuaries are located in urbanized, highly engineered environments. At the same time, they host valuable ecosystems and natural resources. These ecosystems rely on the maintenance of habitat conditions which are constantly changing due to anthropogenic impacts like sea level rise, reservoir operations, and other civil works. This study aims to evaluate the impact of changes in the system forcing on estuarine fine sediment dynamics and budgets. Suspended sediment concentration (SSC) is an important ecosystem health indicator. We apply a process-based modeling approach to assess in detail the impact of climate change and possible engineering scenarios on the San Francisco Bay-Delta system. Model results show that levee breaching may change the entire Delta circulation by increasing the tidal prism and increasing sediment trapping in the Delta. The reduction of sediment input by the watershed has the most impact on habitat conditions while sea level rise decreases mean SSC and trapping efficiency. Our approach shows that validated process-based models are a useful tool to address long-term (decades to centuries) changes in sediment dynamics. In addition, they provide a useful starting point to long-term, process-based studies addressing ecosystem dynamics and health.

This chapter is based on:

Achete, F. M.; van der Wegen, M., Jaffe, B.; Roelvink, D.: How important are climate change and foreseen engineering measures on the sediment dynamics in the San Francisco Bay-Delta system? Submitted to Climatic Changes (submitted)

5.1   Introduction

In estuaries, fine sediments transported from the watershed determine turbidity levels to a high degree. High turbidity levels attenuate light penetration in the water column limiting primary production (Cole et al., 1986), define habitat conditions for endemic species (Brown et al., 2013; Davidson-Arnott et al., 2002), nourish marshes and tidal flats, and carry nutrients and contaminants to intertidal and coastal areas (Feng et al., 1998). Apart from river flow, estuarine turbidity levels depend on land use upstream, tidal energy, and wind-wave resuspension. Natural and anthropogenic changes in environmental conditions modify the sediment dynamics, turbidity levels and the system's resilience to possible future impacts (Hestir et al., 2013; Ibáñez et al., 2014; Schoellhamer, 2011; van Maren et al., 2015; Winterwerp and Wang, 2013).

Estuaries are preferential areas for human settlement and development, hosting urban areas, industries and agriculture (Pont et al., 2002, Douglas and Peltier, 2002, Culliton 1998). The human impact on sediment load is of similar origin for several estuaries worldwide. Initially, an increase in sediment load occurred due to deforestation, changes in land use or mining activities, followed by a sharp decrease of sediment load once river dams were built, and river banks were reinforced (Syvitski and Kettner, 2011; Syvitski et al., 2005; Vörösmarty et al., 2003; Walling and Fang, 2003; Whitney et al., 2014). Examples are the Nile River, the Colorado River, the San Francisco Bay-Delta (Cappiella et al., 1999; Jaffe et al., 1998; Jaffe et al., 2007), the Ebro (Guillén and Palanques, 1997), the Yangtze River Delta (Wang et al., 2015) and Chesapeake Bay (Pasternack et al., 2001). Additional interventions such as channel deepening (van Maren et al., 2015; Winterwerp and Wang, 2013), land reclamation, water deviation for agriculture and human consumption further impact the sediment dynamics (Day et al., 2007; Scavia et al., 2002).

Dating from the Holocene, sea level rise (SLR) has highly influenced estuarine sediment dynamics (Dissanayake et al., 2009; Friedrichs et al., 1990; Sampath et al., 2015; van der Wegen, 2013). SLR is accelerating although it is not uniform worldwide (Nicholls and Cazenave, 2010). Predictions for the State of California are for an increase of between 0.42 and 1.67 m by 2100 (OPC, NRC report). Estuarine vegetation such as marshes, mangroves, and wetlands forest play an important role in erosion control, breeding, refuge and feeding area for wildlife, fishes and invertebrates. They can be resilient to SLR as far as the sediment budget is enough to keep with the rising rate (Fagherazzi et al., 2012; Ibáñez et al., 2014; Kirwan et al., 2010; Reed, 2002; Scavia et al., 2002; Townend et al., 2011).

To maintain prevailing habitat conditions estuaries need to trap sediment under sea level rise scenarios. From the watershed side, a constant decrease in sediments supply is observed since sediment is trapped behind dams. In addition, episodic river floods that once were responsible for resetting the system and transporting large amounts of sediments are now increasingly rare due to flood control barriers and water diversions further decreasing river conveyance capacity. This level of the river management may lead to scenarios of no resetting of the system by floods, depletion of sediment pool downstream and a change from transport limited Suspended Sediment Concentration (SSC) to supply limited SSC (Schoellhamer et al., 2013; Wang et al., 2015).

The aim of this work is to evaluate the possible impact of changes in the estuarine system forcing on sediment dynamics with a focus on the highly engineered estuary of the Sacramento-San Joaquin Delta (Delta). Important disturbances in forcing are change in water diversion operations, levee failure and sea level rise. The main parameters analyzed are sediment budget and spatial and temporal SSC patterns translated into turbidity levels to identify habitat conditions.

Numerical models can assist in defining the tipping points in order to prevent the collapse of the system (Ganju and Schoellhamer, 2010; Wang et al., 2015) and the quantification of human-induced impact increases awareness of possible response of similar systems. We apply a two-dimensional horizontal, averaged in the vertical (2DH), process-based, numerical model (Delft3D FM). The 2DH model solves the 2-D vertically integrated shallow-water equations coupled with advective-diffusive transport and includes sediment transport formulations. (Achete et al., 2015) describe the calibration of the modeled hydrodynamics and SSC levels against observations for the water year 2011. The model thus describes validated sediment dynamics, fluxes and budget in the Delta.

We selected the Delta area as a case study since the area has been well monitored so that detailed model validation can take place. In addition, the system has been highly engineered. There has been an excessive sediment supply by late 19th-century mining activities (Barnard et al., 2015), the marshes were leveed and several dams were built upstream in the Sacramento River and its tributaries.

5.2   Study Area

San Francisco Bay and Sacramento-San Joaquin Delta (Delta) constitutes the San Francisco estuary (Fig 5-1). The entire estuary covers an area of 2900 km² and collects 40% of the total freshwater flow of California discharging in the Delta, seawards through San Francisco Bay and then through Golden Gate being the largest estuary on the US west coast (Jassby et al., 1993; Kimmerer, 2004). The San Francisco estuary hosts several endemic species of fishes and has marshes; moreover, it is densely populated, hosting several industries related to Silicon Valley and intense agribusiness. The different uses culminate in a highly managed estuary with several dams, barriers, bypass for flood control and water diversions for human use and agriculture.

Located landward of San Francisco Bay, the Delta is an inland, natural and man-made, channel network formed by the junction two main tributaries the Sacramento and the San Joaquin River that contributes with 90% and 9% of the discharge respectively, followed by Mokelumne, America and Consumne rivers (Fig 5-1) (Kimmerer, 2004). The tidal wave propagates from San Francisco Bay mouth at Golden Gate, through the embayments of Central Bay, San Pablo Bay, Carquinez Straight, Suisun Bay reaching the Delta output at Chipps Island close to Mallard Island (MAL). MAL experiences mixed diurnal and semidiurnal tide that ranges from about 0.6 m during the weakest neap tides to 1.8 m during the strongest spring tides. During dry season, tidal oscillation is observed in Sacramento River up to the Freeport station (FPT) and in San Joaquin River up to the Vernalis station (VNS).

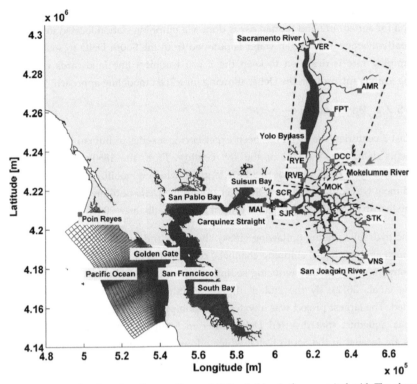

Figure 5-1: Map showing San Francisco Estuary, Bay and Delta. In blue is the numerical grid. The observation stations are represented by the red squares. The Delta areas (North, Central and South) are delimited by the dashed polygons.

The Mediterranean climate dictates the seasonal variability, with the rainfall concentrated in the wet winter months from October to April decreasing towards the summer with September the driest month (Conomos et al., 1985). Water year (WY) is defined from 1 October to 30 September to keep the entire wet season in one hydrological year. The seasons modulate the Delta discharge from low values in dry summer months of 50–150 m^3s^{-1} and to wet spring/winter peak discharges of 800–2500 m^3s^{-1}. Apart from the intra-yearly seasonality the Delta experiences interannual variability with wetter (2011) and drier years (2013).

The Delta is an event driven estuary, during the wet season Sacramento River pulse defines the main sediment dynamics, increasing local SSC and exporting sediment towards the Bay. The hydrological cycle in the Bay–Delta determines the sediment input to the system, and thus biota behavior. During river peak events, large river pulse discharges sediment to the Bay accounting for half of the bay sediment supply (Ganju and Schoellhamer, 2006; McKee et al., 2006; McKee et al., 2013). In average the Delta store two-thirds of the sediment input regardless being a wet and dry year (Schoellhamer et al., 2012; Wright and Schoellhamer, 2005), further changes in the management operations and sea level rise may alter this balance changing the entire ecosystem.

The Delta has several temporary barriers for fish conservancy that is deployed depending on the season; the most important is the Delta Cross Channel (DCC)(Fig 5- 1), which is one of the 3 connections from Sacramento River with the Central and South Delta (Achete, 2015). Water deviation for agriculture and human use is done via pumping station located in the South Delta. On a yearly average 300 m^3s^{-1} of water is pumped from the South Delta to southern California. The pumping rate is designed to keep the 2 psu (salinity) line landwards of Chipps Island avoiding salinity intrusion in the Delta, allowing for a 2DH modeling approach.

5.2.1 Bay history

In the last 3 centuries, the Delta has been experiencing a series of human interventions, the first settlements date from the end of the 18th century. From the 1850`s, the Sacramento River experienced a cycle of 35 years of hydraulic mining, which remobilizes 1.1×10^9 m^3 of sediment. The sediment was transported from the river trough the Delta until the Bay nourishing tidal flats and marshes up to 1m (Jaffe et al., 2007; Wright and Schoellhamer, 2004).

After the hydraulic mining outlawing (1884), the Delta and watershed suffer from intense civil works such as dredging of shipping channels, construction of levee system around the marshes and upstream dams, further reducing sediment input to the Delta and Bay (Delta Atlas, 1995; Whipple et al., 2012). By the end of the 20th-century, major water diversion projects were concluded. The largest project was a water diversion plan consisting of pumping plants and the California aqueduct that diverted Delta water to Southern California. So far, the pumping stations are located in the South Delta, which allows the water and sediment flow through the Delta. Since 1973, the possibility of building a peripheral canal around the Delta to ensure fresh water export to Southern California was discussed. Over the years, the project was modified and today the most possible future extension of the pumping facilities is the project A4, which considers the construction of a tunnel build north from DCC with pumping rate depending on the Sacramento River discharge (http://baydeltaconservationplan.com/Home.aspx).

Between 1957 and 2004, the combination of the ceasing mining and civil works already dropped the Sacramento River sediment supply by 50% (Wright and Schoellhamer, 2004). The sediment load decrease is observed in step changes when major flow events reset the system (Hestir et al., 2013; Schoellhamer, 2011; Schoellhamer et al., 2013). However, with the increasing flood-control projects the reset events are likely to reduce in number or even vanish. Schoellhamer (2011) suggests that the step change is related to crossing the threshold from transport-regulated to supply-limited, which means that the sediment pool is depleted and for this case it is possible to consider for the modeling effort with no initial sediment availability in the bed.

Additional to the water needs the Delta hosts a rich fauna and flora including endemic (e.g. Delta smelt) and endangered (e.g. winter-run salmon) species. In an effort to protect and keep the habitat several temporary barriers for fish conservancy are deployed in the Delta. Recently, there have been discussions of breaching of one or more Delta island to restore marshland, improve habitat and water quality (Suddeth, 2011).

The Delta's history shows how it is engineered. Achete et al. (2015) discussed the sediment dynamics for the WY 2011, and the differences between the wet and dry year. In this work we present a broader analysis of the Delta for 3 WYs (wet, dry and moderate-dry) and we apply the model developed by Achete et. al (2015) to forecast the impacts of further disturbances as pumping activities, levee breaching and the SLR.

5.3 Methodology

5.3.1 Model description

The Delft3D Flexible Mesh (Delft3D FM) (Kernkamp et al., 2010) is a hydro- and morphodynamic unstructured mesh process-based model developed by Deltares (Deltares, 2014). Delft3D FM is a public domain model based on finite volume approach solving shallow-water equations applying a Gaussian solver. Delft3D FM allows for straightforward coupling of its hydrodynamic modules with a water quality model, Delft-WAQ (DELWAQ), which gives flexibility to couple with a habitat (ecological) model (Achete et al., 2015). The detailed formulation and solvers are beyond of the scope of this paper and can be found in (Kernkamp et al., 2010) and downloaded at http://www.D-Flow-baydelta.org/.

Delft3D FM generates time series of the following variables: cell link area; boundary definition; water flow through cell link; pointers that give information about neighboring cells; cell surface area; cell volume; and shear stress file, which is parameterized in Delft3D FM using Manning's coefficient. Based on these processes DELWAQ solves the advection–diffusion–reaction equation for a wide range of substances including fine sediment, the focus of this study. DELWAQ solves sediment source and sink terms by applying the Krone–Parteniades formulation for cohesive sediment transport (Ariathurai and Arulanandan, 1978; Krone, 1962). We applied the 2DH, vertically average results that do not account for stratification. This approach is possible for the Delta because it does not experience salt-fresh water interaction, the pumping operations are such to keep the salinity front (2 psu or X2) seaward from Chipps Island, and the Bay analysis is out of the scope of this work. We also assume that temperature related stratification is limited.

The model grid comprises the full Bay and Delta and it is an updated grid from the described in (Achete et al., 2015). It has average cell size of 1200 m × 1200 m in the coastal area, 450 m × 600 m in the bay area, down to 25 × 25 m in Delta channels (Fig 5-1). All the channels have at least 3 cells describing the cross-section. The present grid has newer bathymetry based on 2009-2011 surveys, includes the Yolo bypass and excludes some dead ends channels in the Delta. It has 74,774 cells and in it takes in an 8 Intel processors computer 3 real days to run 1 year of hydrodynamics simulation and 12 hours for the sediment module. The time step in Delft3D FM is calculated online based in a maximum CFL of 0.7 and it is on average 15 seconds, the sediment module time step is set as 5 minutes.

Both hydrodynamics and sediment calibration parameters settings were established previously by Achete et. al (2015) (http://www.D-Flow-baydelta.org/, Table 5-1). These parameters are applied for all the forecast scenarios.

Table 5-1: Model parameters setting, for hydrodynamics and sediment dynamics.

Model Parameter	Value
Hydrodynamics	
Manning coefficient	0.017 - 0.032 s m$^{-1/3}$
CFL	0.7
Smagorinsky coefficient	0.1
Viscosity	0.1 m^2 s^{-1}
Sediment	
w_s- **fall velocity**	0.5 mm s^{-1}
τ_{cr} - **Critical shear stress**	0.1 Pa
M - erosion coefficient	10^{-5} kg m^{2}s^{-1}

5.3.2 Scenarios

The Bay-Delta has an extensive monitoring network, and, therefore, all the input data to the model are in situ observations. The base-case (BCS) reflects the current conditions and it is used as the standard run for comparison with the other scenarios (Achete et al., 2015). We focus our efforts in the Delta since it hosts several endemic species, it is a highly farmed area, and it is the water source for a large part of the Californian population.

5.3.2.1 Base-Case Scenario -BCS

The seaward Boundary is derived from hourly water level time series at Point Reyes station (tidesandcurrents.noaa.gov/). The station is located at the seaside north of Golden Gate Bridge. The initial water level is 1.5 meter and the model spin up time is less than a week for the 2D simulations.

To represent the variable river flow conditions we modeled 3 WY's, a wet year (2011), a dry year (2012) and one moderate-dry year (2013) (Fig 5-2). Landward, hourly water discharge and SSC boundaries are defined at the main Delta tributaries (http://nwis.waterdata.usgs.gov/nwis and cdec.water.ca.gov/) (Fig 5-1). Two of the 4 landward boundaries are located north of Sacramento to define Sacramento River boundary at Verona (VER) and at American River (AMR) (Fig 5-2). The third boundary is at San Joaquin River at Vernalis (VNS) that together with the Sacramento River account for more that 98% of the Delta discharge (Wright and Schoellhamer, 2005) (Fig 5-2). The last land boundary is at Mokelumne River at Wood Bridge (MOKwb) that discharges in the East Delta.

At VNS and MOKwb stations SSC observation is available and applied as a boundary condition. At VER only the median turbidity in ntu units is available.The turbidity values were converted to SSC by the rating curve (personal communication, USGS Sacramento) $SSC = \exp(0.789 *log10(turb) + 0.567)$, to obtain the boundary condition.The missing values in the time series were approximated by linear interpolation for all the boundaries (Fig 5-2). There are no sediment observations for the American River at AMR. The sediment boundary was defined as a constant concentration of 10mg L^{-1}, similar to the mean concentration of the Sacramento River and

Mokelumne River disregarding the peaks. The discharge of the American River is only 10% of the Sacramento, so such approximation should not impose big errors in the overall sediment flux in the Delta.

Over the model domain, the initial SSC was set at 20 mg L^{-1} because the starting of the simulation is still in the dry season and these are prevailing dry season SSC levels. The model is initialized with no sediment available in the bed following the discussion in Achete et. al (2015) and the findings of Schoellhamer (2011) of a depleted sediment pool. As the sediment pool on the Delta is depleted, considering mud availability in the bottom increases spin up time.

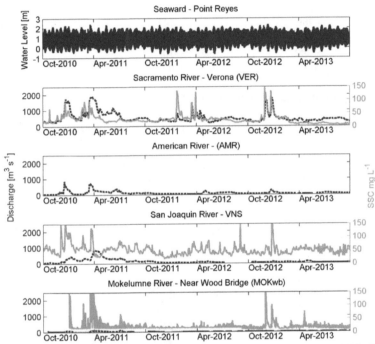

Figure 5-2: Input boundary condition for the 3 WYs, wet (2011), dry (2012) and moderate (2013). Top panel (a) water level at seaward boundary (Point Reyes), the following 4 panels show discharge in dashed blue line and SSC in solid green line for Sacramento River at Verona (b), American River (c), Mokelumne River near Wood Bridge (d) and San Joaquin River at VNS (e) respectively. There are no known data recordings of sediment concentrations for AMR.

In the model, two pumps are implemented at South Delta at Tracy and at Clifton Court. These are the pumps feeding the Southern California aqueduct (http://nwis.waterdata.usgs.gov/nwis). The pumps are defined as negative discharge and will pump sediment depending on local SSC levels.

The model disregards wind-waves because of the Delta geometry. We reasoned that the Delta is mostly a collection of narrow channels without a long enough fetch for a local wind-wave generation. Furthermore, in the flooded islands, the presence of sub-aquatic vegetation protects the bottom sediment from being re-suspended by waves.

5.3.2.2 Pumping Scenario at Sacramento River - SacraP

This scenario applies the seaward and river discharge boundaries of the BCS. The pumping boundaries are shifted from South Delta to the Sacramento River following the BDCP project (Fig 5-3). In the project A4, 3 pumps are planned to be installed upstream from the Delta Cross Channel while pumping operation will depend on the river discharge (Table 5-2) (https://s3.amazonaws.com/californiawater/pdfs/BDCP_FS_Operations.pdf).

Table 5-2: Pumping rate of the project pumping station at Sacramento River, DBCP project a4.

Sacramento Flow (m³ s⁻¹)	Total Pumping (m³ s⁻¹)
> 850	255
> 480	142
> 285	57
> 142	10
<= 142	0

5.3.2.3 Flooded Island Scenario -F-isl

For this case, all the boundaries are kept from the BCS. Flooded island are the former leveed land where the levee breaches forming a "lake". The grid is extended covering Brannan Island, Twitchell Island and Bouldin and Venice Island (Fig 5-3). The levees to be breached are defined based on risk assessment (DWR, risk report 2009), and overtopping probability gives the possible breaching locations (Brooks et al., 2012). Gas mining under Brannan and Twitchell islands increases failure probability. These islands were selected based on the highest probability of flooding in the past and future years. We stress that this is a hypothetical case used as a preliminary assessment on what island flooding could mean for Delta sediment dynamics.

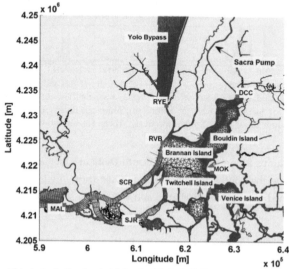

Figure 5-3: Modified grid to include the flooded Island of Twitchell, Brannan, Bouldin and Venice, the breaches are indicated by the red arrows. The black arrow indicates the location of the pumping in Sacrament River from the SacraP scenario.

5.3.2.4 Sea Level Rise scenario at 2100 - SLRS

The sea level rise scenario is defined by incrementing a 1.67 m to the observed Port Reyes water level time-series at the seaward boundary. This value is based on the guideline report by Ocean Protect Centre of California and is considered as a maximum rise in sea level rise after 100 years (http://www.opc.ca.gov/webmaster/ftp/pdf/docs/2013_SLR_Guidance_Update_FINAL1.pdf).

The landward boundaries and grid extend were maintained so that extra flooding of land in the model domain was included. River discharge scenarios suggest a more concentrated wet season earlier snowmelt and a larger rain to snow rate, although the changes are not significant (Knowles and Cayan, 2002). As a result, the average total yearly river hydrograph would keep the same volume.

5.3.2.5 Sea Level Rise scenario at 2100 and decrease of SSC- SLR, 38%SSC

This scenario applies the same seaward and landward water discharge boundary conditions as the in the SLRS. Schoellhamer (2011) show that the SSC load decreased 50% over the past 50 years (about 1.6% decrease per year). Recent data show that the decay rate is less and most probably about 0,8% per year (personal communication, Schoellhamer). This more recent decay rate would lead to a 62% reduction by 2100, and we applied this factor to the SSC landward boundary condition.

5.3.2.6 Sea Level Rise scenario at 2100, Pumping at Sacramento River and Flooded Island - SLRS+SacraP+F-isl

This scenario combines previous scenarios. At the seaward boundary, an increment of 1.67 m is considered (SLRS), Brannan, Twitchell, Bouldin and Venice islands are flooded (F-isl) while the pumping is shifted to Sacramento River (SacraP). In this case, we apply the current SSC condition.

5.4 Results

This section shows the base-case describing the current condition, then the main differences between the scenarios and at last detailed description of each scenario in terms of change in yearly mean water level, location of the contour line delimiting 35 mg L^{-1} concentration in an averaged field over the 3 years, sediment dynamics and budget for the different Delta areas hydrological years.

We divide the Delta in the North, Central, and South to facilitate the analysis. The North Delta is represented by the cross-sections SCR (RVB), and FPT; the South Delta is represented by STK and VNS; the Central Delta is represented by SJR and MOK; the Delta output is at MAL.

5.4.1 Base-Case scenario (BCS)

The isoline of 35 mg L^{-1} corresponds to a turbidity of 18ntu, which is a minimum habitat threshold for an important indicator species, namely the Delta Smelt (Brown et al., 2013). The

Delta Smelt needs high turbidity levels to hide from predators. During the wet year (2011) the isoline defining the exceedance of SSC of 35mg L^{-1} is 12km more seaward compared to the moderate year of 2012, while for the dry year of 2013 the same isoline lays 20 km more landward (Fig 5-4).

In the BCS, the mean water level (MWL) increases towards Sacramento and Vernalis (Fig 5-5), while the tidal wave amplitude is attenuated by bottom friction and river discharge (Fig 5-6). At FPT and VNS, there is little tidal influence and flow is always unidirectional (Fig 5-7). The high discharge events are observed as steps in the cumulative suspended sediment discharge (SSD) (Fig 5-8), and it is possible to track them from FPT downstream in the estuary to MAL.

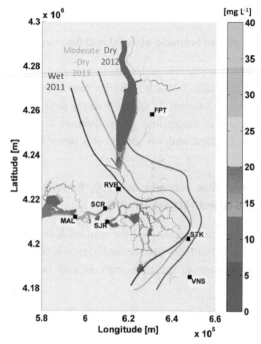

Figure 5-4: BCS mean water year (WY) 2011 SSC map. The color lines show the concentration of 35 mg L^{-1} for each water year. The blue line is the wet WY 2011; the green is dry WY 2012 and the red the moderate-dry WY 2013.

The sediment budget indicates the amount of sediment entering the Delta at FPT and at VNS, the amount leaving at MAL and through the pumps, thus indicating the Delta sediment storage (ΔS). The trapping efficiency (Ψ) is the percentage of sediment storage per amount of sediment supply (Fig 5-9). In WY 2011 ~1,500Kton of sediment entered the Delta, and ~500 Kt was exported resulting in a Ψ of about 60%. Comparing the wet (2011), dry (2012) and moderate-dry (2013) years, we noticed that Ψ is 70%, meaning 10% larger during the dry year than the wet year. Despite the fact that the SSD for WY 2013 is about 30% larger than WY 2011, the Ψ is basically the same.

5.4.2 Scenarios comparison

The scenarios that most affect the MWL are the SLRS and SacraP. The sea level increases the overall MWL and the shift in water pumping location decreases MWL in the North Delta at the same time that increases in the South Delta (Fig 5-5).

The increase of mean sea level in the SLRS decreases bottom friction allowing further penetration of the tidal wave into the Delta, while the flooded islands increase friction dissipating tidal energy, decreasing tidal amplitude (Fig 5-6). The same reasoning applies for the tidal phasing, in the SLRS the tidal wave propagates faster, and the phase is smaller than the BCS while in the F-isl it propagates slower increasing the tidal phase (Fig 5-6).

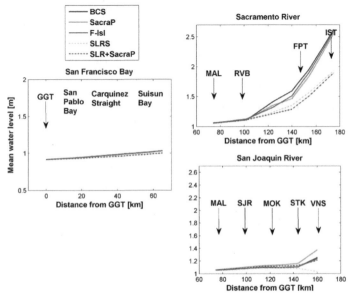

Figure 5-5: WY 2011 mean water level relative to NAVD88 from Golden Gate (GGT) to upstream Sacramento River at IST and upstream San Joaquin River at VNS. The colors represent the scenarios, and the dashed lines are the SLRS and SLRS-P minus the 1.67m extra water level to facilitate the comparison of the water level with the non-SLR scenarios. The arrows indicate locations at the bay and in the Delta.

Changes in the net discharge are less noticeable. All the scenarios lower Sacramento River peak discharge, in favor of a secondary circulation that increases tide residual discharge in the central Delta (Fig 5-7). Despite the small difference in net water discharge, the combination with the changes in MWL and tidal propagation lead to significant impacts on SSD (Fig 5-8). The SLRS increases SSD in the 3 Delta areas (Fig 5-1), while during the F-isl and the SLRS, 38%SSC drastically decrease SSD by about a third in the entire Delta. None of the scenarios affect the sediment transport at VNS. F-isl is the scenario that has the most impact on the trapping efficiency increasing it about 20%. In contrast, SLRS decreases Ψ by 10% (Fig 5-9).

5.4.3 Pumping Scenario - SacraP

Shifting the pumping location impacts the MWL for both rivers (Fig 5-5). In Sacramento River MWL decreases between IST and RVB, while at the San Joaquin River MWL increases until SJR. The change in MWL does not affect tidal propagation in North and Central Delta (Fig 5-6). However, in the South Delta, the higher MWL decreases friction and the tidal amplitude in South Delta are similar to the Central Delta. The pumping does not affect tidal wave phasing.

The pumping upstream in Sacramento River induces recirculation, where discharges in the North Delta are 5% smaller and in Central Delta 5% larger during peak flow. The change in circulation and discharge shift the 35 mg L^{-1} isoline ~5 km landwards (Fig 5-10).

The trapping efficiency in this scenario has some management implications. The project calls for a sedimentation basin before pumping the water southward, but a plan for reintroducing this sediment to the system was not discussed. In this work, we consider that the sediment trapped by the pumps stays in the Delta as part of the trapping efficiency. In this case, pump shifting traps on average 5% more sediment than the BCS.

5.4.4 Flooded Island Scenario - F-isl

Flooding Brannan, Twitchell, Venice and Bouldin Island increases the Delta tidal prism through MAL by 20%. The result of the increase in the tidal prism is a lower mean water level at the Delta (Fig 5-5).The increase of tidal prism attenuates and retards the tidal wave sooner in the Delta (Fig 5-6). Figure 5-6 shows a decrease in amplitude of 30% for the tidal constituents M$_2$ and K$_1$, already at RVB. The island breaching increases the tidal discharge at MAL by 15% but decreases the discharge at RVB by 10% (Fig 5-7).It also retards the discharge peaks at RVB and MOK. The influence of the flooded island does not reach VNS area.

Despite the fact that the flooded island does not affect the sediment flux at FPT and VNS, it creates accommodation space acting as a sediment sink, drastically decreasing the SSD towards the Bay. The mean SSC at Central Delta decreases shifting the isoline line landwards (Fig 5-10). During the dry season, the SSC is halved, and the duration of the high SSC peak decreases. The Ψ increases in more than 20 %, reaching values of 90% for the dry year (Fig 5-9). Further island breaching could lead to an even higher Ψ.

5.4.1 Sea Level Rise scenario at 2100 - SLRS

SLR scenarios can have two possible and counteracting effects on deposition rates. On the one hand, the larger depths and associated smaller flow velocities will lead to a higher trapping efficiency. On the other hand, a larger tidal prism will lead to higher flows and less net deposition (or even erosion). The net result will vary for different case studies.

The increment in mean water level by 1.67 m changes the MWL in the entire Bay-Delta system. To more easily compare the differences in water level in Figure 5-5 the dashed lines show the 2 scenarios considering sea level rise subtracted by 1.67m. The upstream increment of water level is lower for the SLR scenarios. In the bay, the increment remains lower than 0.1 m, but in the Delta, it is more noticeable. At Sacramento River, the difference in water level between IST and MAL is 1.6 m while it is about 0.9 m for the SLR scenario. At San Joaquin River the increment between VNS and MAL for the base scenario is about 0.3 m while the SLR the mean water level increment at VNS is lower than 0.05 m.

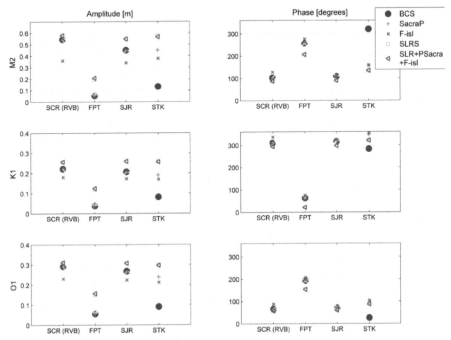

Figure 5-6: Water Level Amplitude at RVB and FPT, SJR and VNS (markers) for all the scenarios, showing the difference in tidal attenuation for amplitude and phase.

The tidal wave amplitude is larger for all the tidal constituents and propagates faster in the SLRS (Fig 5-6). During peak discharges, the peak flow increment the tidal amplitude for the BCS but not in the SLRS where the mean water level is already 1.6 meters higher. The higher MWL leads to less friction and, therefore, larger amplitude. At FPT, the tidal amplitude for the SLRS is ~0.5 m while in the BCS ~0.05m (Fig 5-6).

The discharge reflects the changes in water level, with earlier peaks due to the phase difference, which leads to larger flood and ebb discharges in the whole Delta (Fig 5-6, Fig 5-7). At MAL the difference in discharge is about 50%, increasing landwards especially for the flood discharge. During spring tide, at FPT and STK the tidal energy is still strong enough to keep the discharge bidirectional. San Joaquin branch has lower discharge; therefore, the impact of SLR on the discharge is higher compared to Sacramento River.

The increases in mean sea level and velocities leads to higher mean SSC in the Delta, and the 35 mg L^{-1} is pushed 10 km seawards (Fig 5-10). The deeper water column allows the sediment to be transported further before depositing, increasing SSD in the Delta (Fig 5-8). The sediment being transported farther is reflected in a 10 % trapping efficiency decrease (Fig 5-9). The difference between dry, moderate and wet years follows the same behavior as in the BCS.

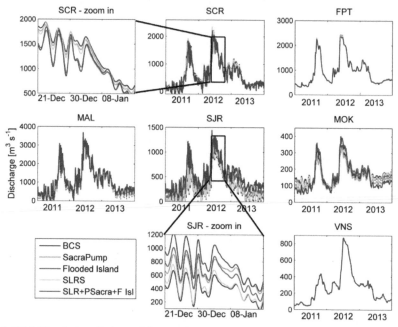

Figure 5-7: Tidal filtered river discharge following Sacramento River at FPT, SCR, and following San Joaquin River, VNS, MOK, and SJR. MAL is the last Delta station and represents the Delta output. The colors represent the scenarios and the zoom in SCR and SJR facilitate the analysis of the difference in discharge during the peak event.

5.4.1 Sea Level Rise scenario at 2100 and decrease of SSC- SLRS, 38%SSC

The hydrodynamics analysis is the same as for the SLRS. The decrease of 62% of the sediment input leads to a sharp decrease in the sediment budget in the entire North and Central Delta (Fig 5-8). It is the only scenario that affects the SSD at FPT. The cumulative SSD drops by one-third in FPT and by half in the other cross-sections.

However, the decrease in sediment input has a minor effect in Ψ. This behavior was already observed in historical observations comparing wet and dry years (Erikson et al., 2013; Wright and Schoellhamer, 2004). The decrease in SSC at the Sacramento boundary leads to a more constant Ψ between dry and wet year, about 65% (Fig 5-9).

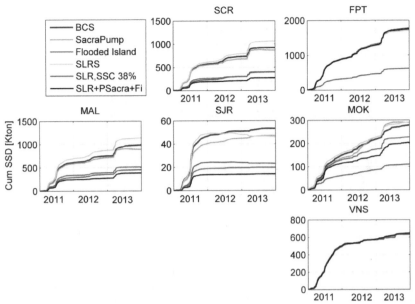

Figure 5-8: Cumulative suspended sediment discharge (SSD) following Sacramento River at FPT, SCR, and following San Joaquin River, VNS, MOK, and SJR. MAL is the last Delta station and represents the Delta output. The colors represent the scenarios. The y-axis limits are different for the different station to highlight the impact of the input change. e.g. MOK cum SSD is one order of magnitude lower than FPT.

5.4.1 Sea Level Rise scenario at 2100 and Pumping at Sacramento River - SLR+SacraP

This scenario is a combination of the SLRS and SacraP, and so it is the hydrodynamics. In Sacramento River, the surplus of MWL is decreased by the change in pumping location (Fig 5-5) and in San Joaquin the MWL increases (Fig 5-5). The change in discharge in this scenario is the same as the changes for the SLRS except at San Joaquin where a minor decrease in discharge is observed.

The sediment dynamics here is a combination of the SLRS and the SacraP. At SCR (RVB) the cumulative sediment transport is 10% lower than the BCS and almost 40% lower than the purely SLRS (Fig 5-9). The mean SSC isoline is between the BCS and the SacraP (Fig 5-10).

5.4.1 Sea Level Rise scenario at 2100, Pumping at Sacramento River and Flooded Island - SLRS+SacraP+Fisl

This scenario is a combination of the SLRS, SacraP, and F-isl. Again the boundaries, FPT, and VNS, are not affected. At Sacramento River, the discharge is similar to the F-isl case only which is lower than the BCS. At San Joaquin River, the combination of changing pumping location, flooded island, and SLR, change the local circulation pattern in the region generating secondary flow, increasing local discharge (Fig 5-7).

Despite larger discharge at San Joaquin River, the cumulative SSD is lower (Fig 5-8).The sediment is trapped in the islands, and at SJR station, the cumulative SSD is one order of magnitude lower than at FPT (Fig 5-9).

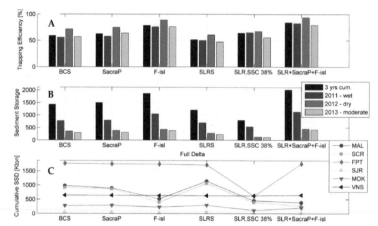

Figure 5-9: a) Trapping efficiency (Ψ) and b) Sediment storage, for each scenario in the x axis, for the full 3 years (black) and for each year, wet (blue), dry (red) and moderate (green). c) Cumulative SSD for the 3 years simulated, for each Delta Station, for each scenario.

5.5 Discussion

5.5.1 Process-based model approach

Process-based models solve the shallow-water and advection-diffusion equations in high spatial and temporal resolutions. The grid defines the spatial resolutions, which are on the of the tens of meters, and the time step, which is on the order of 30 seconds. The high resolution has high demands of computer power, and the models can take from days up to weeks to generate end results. If we were interested only in the overall sediment budget of the system, such a detailed model might not be the best choice. However, we are not only interested in sediment budget but also describing turbidity spatial variation to define habitat condition and deposition pattern. In this context, the detailed model allows generating maps of turbidity distribution in the smaller creeks to study the impact of habitat condition and correlate it with the survival of species. The high spatial/temporal resolution also facilitates the coupling with ecological models and allows the link between abiotic and biotic modeling.

In this work, we do not consider changes in morphology, which may be important for the sea level rise scenarios.

5.5.2 Scenarios sediment budget uncertainties

Defining forecast scenarios result in a number of uncertainties ranging from numerical schemes approximations to the definition of future boundary conditions for each scenario. In this section, we divide the scenarios into 3 groups, namely internal impact, external impact and "non-predictable" impact. The internal impact group comprises the impacts caused by management operation such as changes in pumping operation. In this group uncertainty regarding the new boundaries is rather small as the managers have full control on the forcing.

The external impact group comprises the sea level rise scenarios. SLR prediction is derived from studies considering (IPCC, Stocker et al., 2013) several scenarios of gas emission, eustatic rate, and glacier melt, etc. These studies provide a range of possibilities for sea level rise, which also varies geographically. Among the predictions for the Californian coastline, for the simulations in this work, we choose the worst case scenario of an increase of 1.67 m in the mean sea level (SCR).

The "non-predictable" scenarios are the ones regarding the levees failure. Despite the information about annual breaching (report) and levees elevation (Brooks et al., 2012), information used to choose the most likely islands to fail, the Delta is located in a seismic active region and an expected strong earthquake could lead to a massive levees failure. These are the scenarios with the highest uncertainties and are related to levee failure location and timing while these scenarios are the least likely to be impacted by management decisions to avoid damage.

5.5.3   Internal Impacts - Pumping Scenario

The change in the pumping location from South Delta to Sacramento River changes the sediment path in the Central Delta but has little effect on the Northern Delta. Water discharge in the Central Delta is about one order magnitude smaller than in the North Delta, and the main sediment source is from the Sacramento watershed (Fig 5-9). Southern Delta pumps induce flow from north to south and pump high SSC water from Sacramento River through the Central and South Delta increasing the SSC (Achete et al., 2015). A northward shift in pumping location thus limits SSC levels in the Central Delta as is observed in the SSD at SJR (Fig 5-10).

The decrease in SSC in the Central Delta can affect endemic species survival, such as the Delta Smelt (Brown et al., 2013), and may have an impact on marsh restoration projects. However, the pumping operations are defined by managers and, as such there is the possibility of mitigation of such issues avoiding/changing operations in the Delta.

5.5.4   External Impacts - SLR and SLR with decrease of SSC input

The external impact group consists of simulations with SLR and SLR combined with decreased of SSC input from the watershed. The decrease in SSC is mainly related to dam construction and change in land cover. The biggest difference between SLRS and the BCS is a higher sediment export (Fig 5-8, Fig 5-9), an increase of SSC in the entire Delta (Fig 5-10), but also a decrease in trapping efficiency (dropping by ~10% compared to BCS) (Fig 5-9 a)

The decrease of sediment input from the Sacramento watershed causes the most extreme change in the Delta sediment budget (Fig 5-9 b). As the boundary condition is modified already in FPT the cumulative sediment transport is one-third of the BCS; this condition is observed at RVB, MOK, and SJR, at MAL the transport is halved (Fig 5-9 c). The high rate of SSC decrease is related to the depletion of the sediment pool bayward of the upstream dam construction (Schoellhamer, 2011), so a possible mitigation action to prevent further habitat loss would develop projects for sediment bypassing of dams. This order of sediment input decay would turn marsh restoration projects in the Delta unfeasible (Fig 5-8).

5.5.5 Non-predictable Impacts - Levee Failure

Levee failure is a highly uncertain process due to the probability of earthquake hazard in the region and lack of maintenance plan. The Delta islands have been closed for many decades and currently they are a couple of meters below mean water level. The subsidence has two main reasons: (1) the levees prevent the natural nourishment by river sediment, and (2) the islands are heavily farmed compacting the soil and draining it. The breaching increases the tidal prism by 20%, decreasing tidal amplitude in the Delta (Fig 5-6). The breached islands generate a secondary circulation observed as a phase lag between the high/low water in the channels and inside the islands. The secondary circulation decreases Sacramento River discharge and enhances San Joaquin's. Besides the higher discharge in San Joaquin River, the Ψ is almost 80%, the highest of all the scenarios considered (Fig 5-9).The high Ψ is associated with lower velocities as the river reaches the island in combination with accommodation space. Achete et al. (2015) showed that the breached islands are the main sedimentation area as the river flow reaching the flooded island is not constricted anymore, the velocities drop causing sedimentation of the fine sediment.

The flooded island is the scenario that results in the greatest decrease in the Sacramento River peak discharge at RVB (Fig 5-7), suggesting that in a catastrophic event, if most of the levees fail, the river peak could be fully attenuated inside the Delta. The absence of river peak discharge at the Delta output means no sediment export towards the Bay.

A rather likely scenario in such managed system is a combination of the scenarios of SLR, pumping in Sacramento River and flooded island. In this case, at the downstream stations, the resulting dynamics is similar to the one of decreasing the SSC input (Fig 5-9). Again the Central Delta is the most impacted area, at SJR the SSD is less than a third of the BCS (Fig 5-8, Fig 5-9), during dry years the SSD is so small that almost does not reach the exporting threshold, completely changing the current sediment dynamics in the Delta.

5.5.6 Yearly variability

Apart from the seasonal cycle of the wet and dry season, the Delta experiences interannual cycles of high and low river discharge. The Central Delta suffers the most with the differences in discharge (Fig 5-9 b,c), remembering that the SSD in this area is already one order of magnitude lower than in the North Delta.

In dry and moderate-dry years the Delta budget drops from ~1500 Kt yr^1 (wet) to ~550 Kt yr^1 and the SSD in SJR area is virtually zero (2 Kt yr^1) (Fig 5-8, Fig 5-9). In those years, the influence of SLR, the changing in pumping operation and decrease of SSC input can lead to negative cumulative SSD in Central Delta (Fig 5-8).

Though in dry years the sediment budget is almost half of the wet year sediment budget (Fig 5-9 b), Ψ is on average 10% larger because of the lower velocities the shear stress is lower, and the sediment has a longer time to deposit (Fig 5-9 c). This behavior is observed for all the scenarios.

5.5.7 Ecological Impact

The discussion of ecological impact in the Delta is intrinsically connected to the discussion of natural and anthropogenic impacts. The high ecological value of the Delta including endemic fish species and breeding area for wild salmon aims to keep the system as "natural" as possible. Yet, the Delta is already a highly engineered system with man-made channels, upstream dams (Delta Atlas, 1995), maintenance and dredging of shipping channels, water export and recently, the breaching of levees. In this perspective, all the impacts in the study area are anthropogenic impacts since the area is by itself already man-made.

In another perspective, the Delta still keeps important natural habitat conditions such as turbidity levels higher than 18 ntu (35 mg L^{-1}), which allows for delta smelt survival; sediment trapping in the Delta, facilitating marsh development and avoiding channel erosion.

To compare the scenarios in terms of habitat conditions we define percentage time exceedance of the 35 mg L^{-1} drawing the 10% isoline. Landward (eastward) of the line the SSC exceed 35 mg L^{-1} for more than 10% of the time during the 3 years simulated, the habitat condition is favorable for fishes during more than 10% of the time while seawards the conditions are favorable for less than 10% of the simulation time (Fig 5-10).

Analyzing the Delta by areas the South Delta is not affected by the different scenarios, mainly because the sediment dynamics is defined by the San Joaquin River and it is confined in the South due to low velocities in the channels. This behavior was previously described by Achete at al. (2015).

The Central Delta is the most affected area and closely related to the impacts in the North Delta because it depends on the penetration of the sediment carried by Sacramento River. The decrease of sediment input (SLR, 38%SSC) is the scenario that most affect the habitat condition in the Delta followed by the flooded island (F-isl) and pumping in Sacramento (SacraP) shifting the line north and eastwards. In contrast, SLR improves conditions for the Delta smelt since the sediment stays longer in the water column keeping turbidity higher for a longer period of time. In the SLRS high SSC water flows to Central Delta and all the way to Suisun Bay. The model is 2D, and the SLR will probably lead to salinity intrusion, so the extrapolation to Suisun Bay might be a bit of an overestimation (Fig 5-10). On the contrary, the Delta sediment deposition decreases in the SLRS that is not favorable to marsh restoration.

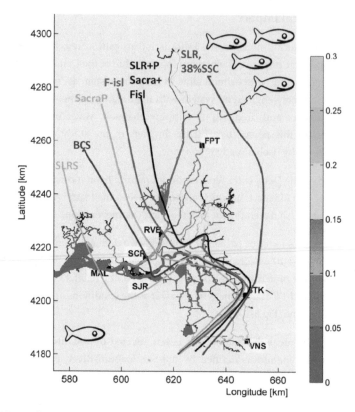

Figure 5-10: BCS exceedance map of SSC higher than 35 mg L⁻¹. The color lines show the 10% exceedance for each scenario. Landward from the line the SSC is higher than 35 mg L⁻¹ during more than 10% of the time and seaward less than 10% of the time.

Recommendations

A better understanding of the river discharge changes in the future scenario could change some of the observed patterns. For example, if the peak drastically decreases all Delta dynamics are likely to change since it is an event driven estuary and instead of export sediment starts to import. As shown, the greatest impact is from a decrease of sediment input. A better description of the river flow could improve the estimate of sediment decrease. A 3D model will help the understanding the salt intrusion and temperature and link to phytoplankton, habitat modeling and the propagation of the sediment plume in the Bay area as well as the import of sediment during the dry season.

5.6 Conclusions

In a highly engineered environment, all impacts are anthropogenic since they modify an already altered system. However, the impacts can improve or deteriorate the current habitat and change system dynamics. The decrease in sediment input is the factor that contributes the most to habitat deterioration as reflected in both sediment budget and SSC distribution. Interestingly, a decrease in sediment input does not significantly affect hydrodynamics and trapping efficiency.

Levee breaching can change the Delta hydrodynamics by increasing the tidal prism and generating secondary circulations and accommodation space. The increase of accommodation space and a decrease in the velocities increases trapping efficiency by 20%. Levee breaching could create a tipping point where instead of exporting sediment the Delta starts to import sediment from the Bay.

Dry years and the Central Delta are the most vulnerable to the system impacts while the wet years and North Delta are most resilient. Regardless of the uncertainty related to the definition of the scenarios, they give indications of the potential impacts allowing for possible remediation. This study and findings for the Sacramento-San Joaquin Delta are a warning signal to many other engineered estuaries worldwide where an increasing number of upstream dams are built disregarding the sediment deficit and impact downstream. We observed several impacts in estuaries. Some of the impacts the by human beings can easily be mitigated through management operations and some not. The outstanding question is, how much should we manage estuaries in order to fit our needs?

The model provides excellent material to future studies on ecosystems in terms of possible changes and mitigation actions. The scenarios here present can now be used as input for ecological models such as phytoplankton, clams, and fishes, enabling to predict the possible impacts.

Acknowledgments
The research is part of the US Geological Survey CASCaDE climate change project (CASCaDE contribution 72). The authors acknowledge the US Geological Survey Priority Ecosystem Studies and CALFED for making this research financially possible. This research was also partially funded by the Brazilian Government via Capes agency. The data used in this work is freely available on the USGS website (nwis.waterdata.usgs.gov). The model applied in this work will be freely available from http://www.D-Flow-baydelta.org/. Rose Martyr provided a helpful review of the manuscript. Lisa Lucas and Noah Knowles supplied invaluable input concerning scenario definition.

6

CONCLUSIONS

6.1 General

In this thesis, I reproduced fine sediment dynamics in the complex geometry San Francisco Bay-Delta estuarine system using the flexible mesh, process-based model, Delft3D FM. Based on solving the hydrodynamic shallow water equations and the advection-diffusion equation for sediment, the model allows for describing the spatial and temporal variability of suspended sediment concentration (SSC), which is subsequently translated into turbidity levels, deposition patterns, sediment budgets and coupling with ecological and contaminants models. We applied the model at a long-term, system scale as well as a seasonal, local slough scale. The main forcing determining the sediment dynamics distinguishes these two scales. Temporal, high river flow is the main sediment dynamic driver in the Sacramento-San Joaquin Delta, which is called here an event-driven estuary. The Alviso Slough main sediment driver is tidal forcing, therefore, referred to as a tide-driven estuary.

Regarding bed sediment availability, we can divide estuaries into two groups. One has plenty of sediment available at the bottom (rich sediment pool). In this case, the transport capacity determines the SSC. The other group has little sediment available at the bottom (depleted sediment pool) so that the SSC is determined by the sediment entering from the boundaries. Many tide-driven estuaries (e.g. Rhine, Ems, Alviso Slough and Loire) have a large sediment pool because they have two sediment sources, i.e. riverine and coastal On the other hand, event-driven estuaries (e.g. the Delta) do not necessarily have a depleted sediment pool. In the Delta, the sediment pool depletion results from the river damming that decreases sediment supply and enhances erosion of the river bed.

Ongoing discussion in academia is about the applicability of research and bridging the gap between different areas of sciences and between science and society. In this context, this thesis has three dimensions: a scientific, a multidisciplinary and a managerial one. The scientific dimensions' objective is to investigate the fine sediment dynamics at a variety of spatial and temporal scales for both an event- and a tide-driven estuary. Concerning multidisciplinarity, this work is embedded in the multidisciplinary CASCaDE II project (Chapter 1). The Bay-Delta model (described in this thesis) is one of the first real case applications of the new numerical model (Delft3D FM), which required much interaction with the code developers. I received input from the upstream groups (climate and boundary conditions), and the results of this work (SSC) is used as input for ecological and contaminant modeling. Management wise, this research provides a calibrated model to understand the system, to assist in the Delta operations, surveying campaigns and further scenarios simulations.

The main scientific objectives are answered following the research questions presented in chapter 1.

Is a large scale, process-based model a suitable tool for reproducing sediment dynamics and sediment budget in complex geometry estuaries?

Yes. In this thesis (Chapters 2, 3 and 4) I show that the 2D process-based model Delft3D FM, after limited calibration, can accurately reproduce fine sediment dynamics in terms of suspended sediment concentration (turbidity levels) and a detailed, yearly sediment budget. In this thesis, I consider "large scale" because the model domain encompasses the entire estuarine region, from the river to the coast area (spatial resolution of O (10-100m)), and a timeframe that covers a yearly scale (with seconds to minute time steps). Chapter 2 shows a successful application to the event-driven Sacramento-San Joaquin Delta while Chapter 4 shows a tide-driven application of Alviso Slough. The models were calibrated for intertidal dynamics, and the two applications give the confidence to reproduce spatial maps of fine sediment dynamics ranging from hourly time scale to yearly budgets.

Sediment budgets are often derived from in situ observations, which require an extensive observation network to describe the hydro-sediment dynamics in time and space. For many estuaries worldwide, this network is non-existent, and/or data is not available. Numerical models such as the process-based model (applied in this study) data-based or behavior-oriented models (empirical models) are used to simulate sediment dynamics. Empirical models are suitable for calculating general sediment budget, but they have a too coarse resolution to solve detailed spatial and temporal scales and often lack a physical explanation. Also, the empirical relationships may not be valid for future conditions.

Process-based models are known to be complex and computationally expensive by including many processes, but can reach a much higher resolution in time and space. Solving the physical equations enables to study the main physical forcings, the sensitivity of the system to parameter change and to forecast impacts in the estuarine dynamics. The high spatial and temporal resolution of the process-based model results are important to define spatial SSC distribution, erosion/deposition patterns and to couple abiotic (hydrodynamics and sediment dynamic) with biotic (phytoplankton, benthos, and fishes) models. Process-based models are able to substantiate data derived sediment budgets and other sediment dynamics by providing physical explanations for observed phenomena and providing a physically sound basis for possible future developments.

Process-based models demand high computational power to solve the physical equations. It is important to be able to define the main processes to simplify and interpret the model results. A large validation dataset enables us to investigate the minimum relevant input to calibrate and validate the model.

How much in situ data is necessary to develop a calibrated sediment model for complex geometry estuaries?

A large dataset is helpful to understand the system dynamics, but not all the data is necessary to set and calibrate the model. This thesis shows that with simple sediment settings of one fraction at the input boundary and a simple distribution of bed sediment availability, it is possible to reproduce seasonal variations as well as construct a yearly sediment budget with more than 90% accuracy when compared with a data derived budget (Chapters 2 and 4). The Delta is a test case where the in situ observations are abundant. This allows exploring the amount of data needed to develop an estuarine sediment budget model.

A robust event-driven estuarine model calls for reliable bathymetry, water discharge and suspended sediment concentration (SSC) at the landward boundaries and water level at the seaward boundary (Chapters 2, 3 and 4). For the tide-driven estuary, SSC is needed at the seaward boundary as well (Chapter 4). It is necessary to have in situ observations close to the estuary output for the event-driven estuary (Chapter 2) or in the middle of the estuary for the tide-driven (Chapter 4), to calibrate and validate the model. With this simple setting and considering one mud fraction it is possible to reproduce sediment dynamics of less-measured estuaries.

Alviso Slough (tide-driven estuary) has a large sediment pool. To reach the high SSC observed in the slough, we defined two meters of one sediment fraction available over the entire domain. The Delta (event-driven estuary) has a depleted sediment pool. In this case, it was possible to define no bed sediment availability, since the main source of sediment is the river input.

In case that the type of estuary is not known, it is advised to start with a simple distribution of one fraction of sediment over the domain. If the estuary is a sediment depleted estuary, it will take considerable time to wash away the bed sediment (for example, one month of simulation time). If the estuary has a high sediment pool, the initial condition will provide enough sediment to reproduce the required high SSC within some tidal cycles.

What do we learn from comparing time/spatial scales of an event-driven system and a tide-driven system?

In this thesis I investigated the sediment dynamics in a tide- and in an event-driven system, each one having distinct spatial/temporal scales (table 6-1). The results from this thesis lead us to propose a classification into event- and tide-driven estuaries, considering the main forcing to net suspended sediment transport.

Table 6-1: Comparison between main characteristics of a tidal and an event-driven estuary.

	Event-driven estuary (Delta)	Tide-driven Estuary (Alviso)
Main Sediment Forcing	River Discharge	Tides
SSC timescale	Days - weeks	Hours
Morphodynamic adaptation time scale	Weeks - months	Years
Boundary	Landward	Seaward
Main sediment calibration parameter	Fall Velocity	Erosion Coefficient
Main sediment transport direction	Unidirectional - watershed towards the Bay	Bidirectional

This classification unfolds in 4 possibilities: large and small scale event-driven estuaries and large and small scale tide-driven estuaries. San Francisco estuary provides two test cases: Sacramento-San Joaquin Delta as a large scale event-driven estuary with a spatial scale in the order of 50 km and temporal scale in the order of days/weeks; and the Alviso Slough as a small scale tidal-driven estuary with a spatial scale in the order of 10km and temporal scale in the order of hours/days.

We show the particularities of each system (Chapter 2, 3 and 4) and thereafter draw the parallel between them (Table 6-1).

The Delta is an event-driven estuary because the main sediment driver is the seasonal river peak events that transport sediment from the watershed towards the Bay (Chapter 2). An accurate definition of the river boundary condition is mandatory. In the Delta, tides have a minor effect on net sediment transport. The tides stir sediment to make it available for river transport (during high river flow) and rework the sediment previously deposited by the river load (with limited impact, during low river flow conditions) (Chapter 3). The river events occur during the wet season lasting for about 2-3 months, with higher peaks on a weekly time scale. In this case, we are able to reproduce turbidity levels and budget using tidally averaged results reducing the amount of data to be processed. The Delta has sandy channels and the most important sediment parameter is the fall velocity. Event-driven estuaries have morphodynamic adaptation time scales in the order of weeks to months (this is the period during which largest dynamics occur), and the sediment carried by the river is transported over tens of kilometers.

The Alviso Slough is a tide-driven estuary. Thus, the main sediment driver is the tidal wave penetrating into the estuary. Occasionally, river floods generate a unidirectional slough sediment flux delivering large amounts of sediment to the slough and bay. For a short period, the estuary becomes event-driven. After the river event, the tide becomes the main forcing again, and the balance between a tidal wave and river discharge redefines the sediment flux. This balance results in erosion and deposition areas within the slough. In this case, the seaward boundary condition is essential for an accurate model. As the sediment is resuspended every tidal cycle, the erosion coefficient is the most important sediment parameter and will dictate how much sediment is picked up by each tidal cycle. The SSC levels follow the tidal cycle, during flood and ebb SSC is higher and during slack-water, SSC is lower. The tidal time scale is in the order of hours, and the asymmetry of this cycle will define the net sediment transport. The sediment is transported by a few kilometers following the tidal excursion. The morphodynamic adaptation time scale is long due to the low, tide-residual sediment transports.

The two systems behave differently due to a combination of forcing and spatial/temporal scales. Event-driven estuaries are simpler because the rivers are the main sediment source leading to the unidirectional net sediment transport throughout the estuary. The ratio between the estuary spatial scale and sediment fall velocity will determine the sediment deposition in the estuary and availability for the dry season. In the tide-driven estuary, the sediment pool is the main sediment source and river input does not play a major role over the long term. In this case, the tidal phase calibration is very important to the sediment transport.

Despite the differences in forcing and spatial/temporal scales between the tide- and event-driven estuaries, the processes governing both estuarine type dynamics are covered by our process-based approach and so are the prescribed parameter settings. The differences in the modeling are the parameters calibration values and the sensitive parameters (Table 6-1).

To what extent can we predict future scenarios? What is the model applicability?

The model presented in this thesis was applied to forecast possible management, engineering and sea level rise impacts. In an event-driven estuary where the sediment pool is poor, and considering that the changes in morphology are less important, we can simulate long-term sea level rise scenarios considering the current bathymetry (Chapter 4).

In a tide-driven estuary, where the sediment pool is rich, breaching of levees intensifies erosion at the mouth and close to the breaches. After some time, the estuary tends to approach a new equilibrium with a decreasing the erosion rate. Without a morphological model, we can calculate the short-term (up to six months-one year) transient sediment dynamics based on empirical relationships. However, long-term forecast requires feedback between hydrodynamics and morphology (Chapter 4).

In both cases, we observe that engineering works have been changing estuaries in a more profound and faster way than any natural process (Chapter 4) (Cloern et al., 2015). Alviso Slough, a tide-driven estuary, is vulnerable to changes tidal prism (e.g. levee failure and marshes restoration) (Chapter 4). The geometry influences the tidal propagation and asymmetry, which are the main sediment drivers. It is possible that this finding applies for most of the tide-driven estuaries. Event-driven estuaries are less vulnerable to changes in geometry (Chapter 03) and to downstream changes (e.g. sea level rise) (Chapter 3). However, changes in the upstream boundaries (e.g. river damming and flood control) have a high impact in the system sediment dynamics (Chapter 5).

Forecasting is associated with uncertainty. The process-based model has uncertainties related to the numerical scheme approximations and numerical instabilities. Also, uncertainties are related to the model parameter settings (e.g. forcing and sediment characteristics). The definition of scenarios boundary conditions has uncertainties regarding the future sea level, changes in hydrological cycle, future management operations and levees failure. Despite all the uncertainties, forecast scenarios indicate the possible trends of the system.

Sediment is a key factor in the water quality and ecology of an estuary. The impacts can improve or deteriorate the current habitat conditions. The Delft3D FM software allows direct coupling with water quality, sediment transport and habitat models (Chapter 2 and 4). Our work provides the basis for a chain of models, from the hydrodynamics, to suspended sediment transport and ecology (phytoplankton, fish, clams, and marshes). The turbidity and deposition pattern analysis may guide ecologists in future works to define areas of interest and/or vulnerable areas to be studied, as well as guide data collection efforts.

6.2 Recommendations for future research

-   **Three-dimensional (3D) modeling approach**

The grid flexibility allows for a combined modeling of the Bay and Delta system. In this thesis, we were interested in the Delta and applied a two-dimensional vertically-integrated approach.

However, the Bay experiences salt front intrusion and to expand the analysis to the Bay and coastal area a 3D approach is required.

- **Morphodynamics implementation.**

Offline coupling enables faster calibration, defining turbidity levels and calculating a budget. However, offline coupling does not allow calculating morphological changes and the feedback between morphology and hydrology. The Delft3D FM is under development and implementation of the morphological update will expand the applicability of long-term simulations especially in the tidal-driven environment.

- **Considering climate changes impacts**

Climate changes are not restricted to sea level rise, despite being the only parameter considered in this thesis. IPCC studies show changes in temperature, intensity and periodicity of precipitation and snow melt. These factors may affect the hydro-sediment dynamics of estuaries as well. A better understanding of these changes would improve the forecast scenarios and give more room for the system adaptability.

- **Vegetation**

Vegetation including marshland, mangrove, and sub-aquatic vegetation play an important role in the sediment dynamics and are found in most estuaries. Vegetation traps and fixes sediment working as a protection layer for sediment on the bed, banks, and shore. Implementation of vegetation module with feedback to the hydro-morphodynamic model would enhance long-term predictions and add to the completeness of estuarine models.

- **Further management scenarios**

In this thesis, I presented some managerial scenarios including changes in pumping location and decrease of sediment input due to river damming. However, there are several options for pumping operation, temporary barriers, marsh restoration, levee maintenance or breaching that were not considered because of time constraints.

- **Social impact**

This research has a high impact on people living in the Delta area. The results can be presented in the form of brochure, animations and infographic to reach a broader audience bringing awareness of the environment.

7

REFERENCES

Achete, F.M., van der Wegen, M., Roelvink, D., Jaffe, B., 2015a. A 2-D process-based model for suspended sediment dynamics: a first step towards ecological modeling. Hydrology and Earth System Sciences 19, 2837-2857.

Achete, F. M.; van der Wegen, M.; Roelvink, D., Jaffe, B., 2016: Suspended Sediment Dynamics in a tidal channel network under Peak River Flow, Ocean Dynamics DOI: 10.1007/s10236-016-0944-0

Ackerman, J.T., Eagles-Smith, C.A., 2010. Agricultural Wetlands as Potential Hotspots for Mercury Bioaccumulation: Experimental Evidence Using Caged Fish. Environmental Science & Technology 44, 1451-1457.

Ackerman, J.T., Marvin-DiPasquale, M., Slotton, D., Eagles-Smith, C.A., Herzog, M.P., Hartman, C.A., Agee, J.L., Ayers, S., 2013. The South Bay Mercury Project: Using Biosentinels to Monitor Effects of Wetland Restoration for the South Bay Salt Pond Restoration Project, The South Bay Mercury Project: Using Biosentinels to Monitor Effects of Wetland Restoration for the South Bay Salt Pond Restoration Project. USGS, p. 227p.

Ariathurai, R., Arulanandan, K., 1978. Erosion Rates of Cohesive Soils. Journal of the Hydraulics Division 104, 5.

ASTM International, 2002. Standards on Disc.

Aubrey, D.G., 2013. Hydrodynamic Controls on Sediment Transport in Well-Mixed Bays and Estuaries, Physics of Shallow Estuaries and Bays. Springer-Verlag, pp. 245-258.

Barnard, P.L., Schoellhamer, D.H., Jaffe, B.E., McKee, L.J., 2013. Sediment transport in the San Francisco Bay Coastal System: An overview. Marine Geology 345, 3-17.

Baskerville, B., Lindberg, C., 2004. The Effect of Light I ntensity, Alga Concentration, and Prey Density on the Feeding Behavior of Delta Smelt Larvae, American Fisheries Society Symposium. Citeseer, pp. 219-227.

Bergamaschi, B., Kuivila, K., Fram, M., 2001. Pesticides associated with suspended sediemnts entering San Francisco Bay following the first major storm of water year 1996. Estuaries 24, 368-380.

Bertrand-Krajewski, J.-L., Chebbo, G., Saget, A., 1998. Distribution of pollutant mass vs volume in stormwater discharges and the first flush phenomenon. Water Research 32, 2341-2356.

Bever, A.J., MacWilliams, M.L., 2013. Simulating sediment transport processes in San Pablo Bay using coupled hydrodynamic, wave, and sediment transport models. Marine Geology.

Boesch, D.F., Brinsfield, R.B., Magnien, R.E., 2001. Chesapeake bay eutrophication. Journal of Environmental Quality 30, 303-320.

Boumans, R.M.J., Burdick, D.M., Dionne, M., 2002. Modeling Habitat Change in Salt Marshes After Tidal Restoration. Restoration Ecology 10, 543-555.

Brennan, M.L., Schoellhamer, D.H., Burau, J.R., Monismith, S.G., 2002. Tidal asymmetry and variability of bed shear stress and sediment bed flux at a site in San Francisco Bay, USA. Environmental Fluid Mechanics Laboratory, Dept. Civil & Environmental Engineering, Stanford University, Stanford, CA,

U. S. Geological Survey, Placer Hall, 6000 J St., Sacramento, CA p. 15.

Bricker, J., 2003. Bed drag coefficient variability under wind waves in a tidal estuary: field measurements and numerical modeling, Civil and Environmental Engineering. Stanford University, Stanford, CA, p. 161.

Brooks, B.A., Bawden, G., Manjunath, D., Werner, C., Knowles, N., Foster, J., Dudas, J., Cayan, D., 2012. Contemporaneous Subsidence and Levee Overtopping Potential, Sacramento-San Joaquin Delta, California. San Francisco Estuary and Watershed Science 10.

Brown, L.R., 2003. A Summary of the San Francisco Tidal Wetlands Restoration Series. San Francisco Estuary and Watershed Science 1.

Brown, L.R., Bennett, W.A., Wagner, R.W., Morgan-King, T., Knowles, N., Feyrer, F., Schoellhamer, D.H., Stacey, M.T., Dettinger, M., 2013. Implications for Future Survival of Delta Smelt from Four Climate Change Scenarios for the Sacramento-San Joaquin Delta, California. Estuaries and Coasts 36, 754-774.

Brunn, P., Chiu, T., Gerritsen, F., Morgan, W., 1962. Storm tides in Florida as Related to coastal topography. Eng. Progr. at Univ. Florida Eng. Ind. Exp. Station, Gainesville, Florida.

Cai, H., Savenije, H.H.G., Toffolon, M., 2014. Linking the river to the estuary: influence of river discharge on tidal damping. Hydrol. Earth Syst. Sci. 18, 287-304.

Callaway, J.C., Parker, V.T., Vasey, M.C., Schile, L.M., Herbert, E.R., 2011. Tidal Wetland Restoration in San Francisco Bay: History and Current Issues. San Francisco Estuary and Watershed Science 9.

Callaway, J.C., Schile, L.M., Borgnis, E.L., Busnardo, M., Archbald, G., Duke, R., 2013. Sediment Dynamics and Vegetation Recruitment in Newly Restored Salt Ponds: Final Report for Pond A6 Sediment Study South San Francisco Bay Salt Pond Restoration Projec, Sacramento, p. 29.

Capo, S., Brenon, I., Sottolichio, A., Castaing, P., Le Goulven, P., 2009. Tidal sediment transport versus freshwater flood events in the Konkouré Estuary, Republic of Guinea. Journal of African Earth Sciences 55, 52-57.

Capobianco, M., DeVriend, H.J., Nicholls, R.J., Stive, M.J., 1999. Coastal area impact and vulnerability assessment: the point of view of a morphodynamic modeller. Journal of Coastal Research, 701-716.

Cappiella, K., Malzone, C., Smith, R., Jaffe, B.E., 1999. Sedimentation and Bathymetry changes in Suisun Bay: 1867-1990, in: USGS (Ed.). USGS Menlo Park.

Casulli, V., Walters, R.A., 2000. An unstructured grid, three-dimensional model based on the shallow water equations. International Journal for Numerical Methods in Fluids 32, 331-348.

Cloern, J.E., Abreu, P.C., Carstensen, J., Chauvaud, L., Elmgren, R., Grall, J., Greening, H., Johansson, J.O.R., Kahru, M., Sherwood, E.T., Xu, J., Yin, K., 2015. Human activities and climate variability drive fast-paced change across the world's estuarine–coastal ecosystems. Global Change Biology, n/a-n/a.

Coco, G., Murray, A.B., Green, M.O., Thieler, E.R., Hume, T.M., 2007. Sorted bed forms as self-organized patterns: 2. Complex forcing scenarios. Journal of Geophysical Research 112.

Cole, B., Cloern, J., Alpine, A., 1986. Biomass and productivity of three phytoplankton size classes in San Francisco Bay. Estuaries 9, 117-126.

Conomos, T.J., Smith, R.E., Gartner, J.W., 1985. Environmental setting of San Francisco Bay. Hydrobiologia 129, 12.

Dastgheib, A., Roelvink, J.A., Wang, Z.B., 2008. Long-term process-based morphological modeling of the Marsdiep Tidal Basin. Marine Geology 256, 90-100.

Davidson-Arnott, R.G.D., van Proosdij, D., Ollerhead, J., Schostak, L., 2002. Hydrodynamics and sedimentation in salt marshes: examples from a macrotidal marsh, Bay of Fundy. Geomorphology 48, 209-231.

Day, J.W., Gunn, J.D., Folan, W.J., Yáñez-Arancibia, A., Horton, B.P., 2007. Emergence of complex societies after sea level stabilized. Eos, Transactions American Geophysical Union 88, 169-170.

De Vriend, H.J., 1991. Mathematical modelling and large-scale coastal behaviour. Journal of Hydraulic Research 29, 741-753.

De Vriend, H.J., Ribberink, J., 1996. Mathematical modelling of meso-tidal barrier island coasts. Part II: Process-based simulation models. Advances in coastal and ocean engineering 2, 151-198.

Deletic, A., 1998. The first flush load of urban surface runoff. Water Research 32, 2462-2470.

Delta Atlas, 1995. Sacramento-San Joaquin Delta Atlas. DWR - Department of Water Resources.

Deltares, 2014. D-Flow Flexible Mesh, Technical Reference Manual. Deltares, Delft, p. 82.

Dissanayake, D.M.P.K., Roelvink, J.A., van der Wegen, M., 2009. Modelled channel patterns in a schematized tidal inlet. Coastal Engineering 56, 1069-1083.

Dissanayake, S., Zhao, Y., Sugahara, S., Takenaka, M., Takagi, S., 2011. Channel direction, effective field, and temperature dependencies of hole mobility in (110)-oriented Ge-on-insulator p-channel metal-oxide-semiconductor field-effect transistors fabricated by Ge condensation technique. Journal of Applied Physics 109, 033709.

Downing-Kunz, M., Schoellhamer, D., 2015. Suspended-Sediment Trapping in the Tidal Reach of an Estuarine Tributary Channel. Estuaries and Coasts, 1-15.

Downing, J., 2006. Twenty-five years with OBS sensors: The good, the bad, and the ugly. Continental Shelf Research 26, 2299-2318.

Dronkers, J., 1986. Tidal asymmetry and estuarine morphology. Netherlands Journal of Sea Research 20, 117-131.

Dyer, K., 1986. Coastal and estuarine sediment dynamics. Chichester: Wiley.

Dyer, K.R., Christie, M.C., Feates, N., Fennessy, M.J., Pejrup, M., van der Lee, W., 2000. An Investigation into Processes Influencing the Morphodynamics of an Intertidal Mudflat, the Dollard Estuary, The Netherlands: I. Hydrodynamics and Suspended Sediment. Estuarine, Coastal and Shelf Science 50, 607-625.

Edwards, K.P., Barciela, R., Butenschön, M., 2012. Validation of the NEMO-ERSEM operational ecosystem model for the North West European Continental Shelf. Ocean Sci. 8, 983-1000.

Elias, E.P.L., van der Spek, A.J.F., 2006. Long-term morphodynamic evolution of Texel Inlet and its ebb-tidal delta (The Netherlands). Marine Geology 225, 5-21.

Erikson, L.H., Wright, S.A., Elias, E., Hanes, D.M., Schoellhamer, D.H., Largier, J., 2013. The use of modeling and suspended sediment concentration measurements for quantifying net suspended sediment transport through a large tidally dominated inlet. Marine Geology.

Fagherazzi, S., Kirwan, M.L., Mudd, S.M., Guntenspergen, G.R., Temmerman, S., D'Alpaos, A., van de Koppel, J., Rybczyk, J.M., Reyes, E., Craft, C., Clough, J., 2012. Numerical models of salt marsh evolution: Ecological, geomorphic, and climatic factors. Reviews of Geophysics 50, RG1002.

Feng, H., Kirk Cochran, J., Lwiza, H., Brownawell, B.J., Hirschberg, D.J., 1998. Distribution of heavy metal and PCB contaminants in the sediments of an urban estuary: The Hudson River. Marine Environmental Research 45, 69-88.

Feng, S., Ai, Z., Zheng, S., Gu, B., Li, Y., 2014. Effects of Dryout and Inflow Water Quality on Mercury Methylation in a Constructed Wetland. Water, Air, & Soil Pollution 225, 1-11.

Foxgrover, A.C., Finlayson, D.P., Jaffe, B., Fregoso, T.A., 2011. Bathymetry and Digital Elevation Models of Coyote Creek and Alviso Slough, South San Francisco Bay, California, Open-File Report. U.S. Geological Survey p. 23.

Foxgrover, B.J.A., 2007. Sediment Deposition and Erosion in South San Francisco Bay,California from 1956 to 2005.

Fraysse, M., Pinazo, C., Faure, V.M., Fuchs, R., Lazzari, P., Raimbault, P., 2013. Development of a 3D Coupled Physical-Biogeochemical Model for the Marseille Coastal Area (NW Mediterranean Sea): What Complexity Is Required in the Coastal Zone? PLoS One 8.

Fregoso, T.A., Foxgrover, A.C., Jaffe, B.E., DiPasquale, M.M., 2014. Using bathymetric surveys to estimate mercury mobilization from scour within Alviso Slough 8th Biennial Bay-Delta Science Conference, Sacramento, Califronia.

Friedrichs, C., Aubrey, D., 1988. Non-linear tidal distortion in shallow well-mixed estuaries: a synthesis. Estuarine Coastal and Shelf Science 27, 521-545.

Friedrichs, C.T., 2010. Barotropic tides in channelized estuariesContemporary Issues in Estuarine Physics. Cambridge University Press.

Friedrichs, C.T., Aubrey, D.G., Speer, P.E., 1990. Impacts of relative sea-level rise on evolution of shallow estuaries, Residual currents and long-term transport. Springer, pp. 105-122.

Gallo, M., Vinzon, S., 2005. Generation of overtides and compound tides in Amazon estuary. Ocean Dynamics 55, 441-448.

Ganju, N.K., Schoellhamer, D.H., 2006. Annual sediment flux estimates in a tidal strait using surrogate measurements. Estuarine, Coastal and Shelf Science 69, 165-178.

Ganju, N.K., Schoellhamer, D.H., 2009a. Calibration of an estuarine sediment transport model to sediment fluxes as an intermediate step for simulation of geomorphic evolution. Continental Shelf Research 29, 148-158.

Ganju, N.K., Schoellhamer, D.H., 2009b. Decadal-Timescale Estuarine Geomorphic Change Under Future Scenarios of Climate and Sediment Supply. Estuaries and Coasts 33, 15-29.

Ganju, N.K., Schoellhamer, D.H., 2010. Decadal-timescale estuarine geomorphic change under future scenarios of climate and sediment supply. Estuaries and Coasts 33, 15-29.

Ganju, N.K., Schoellhamer, D.H., Warner, J.C., Barad, M.F., Schladow, S.G., 2004. Tidal oscillation of sediment between a river and a bay: a conceptual model. Estuarine, Coastal and Shelf Science 60, 81-90.

Gibbs, R.J., Wolanski, E., 1992. The effect of flocs on optical backscattering measurements of suspended material concentration. Marine Geology 107, 289-291.

Gilbert, G.K., 1917. Hydraulic-mining debris in the Sierra Nevada, in: Paper, P. (Ed.), USGS Numbered Series. USGS, p. 154.

Goodwin, P., Denton, R.A., 1991. TECHNICAL NOTE. SEASONAL INFLUENCES ON THE SEDIMENT TRANSPORT CHARACTERISTICS OF THE SACRAMENTO RIVER, CALIFORNIA, ICE, 1 ed. ICE, pp. 163-172.

Grant, W.D., Madsen, O.S., 1979. Combined wave and current interaction with a rough bottom. Journal of Geophysical Research: Oceans 84, 1797-1808.

Guillén, J., Palanques, A., 1997. A historical perspective of the morphological evolution in the lower Ebro river. Environmental Geology 30, 174-180.

Guo, L., van der Wegen, M., Roelvink, J.A., He, Q., 2014. The role of river flow and tidal asymmetry on 1-D estuarine morphodynamics. Journal of Geophysical Research: Earth Surface 119, 2014JF003110.

H. T. Harvey & Associates, Philip Williams and Associates, Associates, M., Staff of San Francisco Bay Conservation and Development Commision, 1982. Guidelines for Enhancement and Restoration of Diked Historic Baylands. Technical Report prepared for the San Francisco Bay Conservation and Development Commission, April 1982, p. 100 plus appendices.

Hayes, T., John, K.J.R., Wheeler, N.J.M., 1984. California Surface Wind Climatology. Aerometic Projects and Laboratory Branch, p. 180.

Hervouet, J.-M., 2007. Hydrodynamics of free surface flows: modelling with the finite element method. John Wiley & Sons.

Hestir, E.L., Schoellhamer, D.H., Morgan-King, T., Ustin, S.L., 2013. A step decrease in sediment concentration in a highly modified tidal river delta following the 1983 El Nino floods. Marine Geology 345, 304-313.

Hoitink, A.J.F., Hoekstra, P., van Maren, D.S., 2003. Flow asymmetry associated with astronomical tides: Implications for the residual transport of sediment. Journal of Geophysical Research: Oceans 108, 3315.

Horrevoets, A.C., Savenije, H.H.G., Schuurman, J.N., Graas, S., 2004. The influence of river discharge on tidal damping in alluvial estuaries. Journal of Hydrology 294, 213-228.

http://southbayrestoration.org/.

Hu, J., Li, S., Geng, B., 2011. Modeling the mass flux budgets of water and suspended sediments for the river network and estuary in the Pearl River Delta, China. Journal of Marine Systems 88, 252-266.

Ibáñez, C., Day, J.W., Reyes, E., 2014. The response of deltas to sea-level rise: Natural mechanisms and management options to adapt to high-end scenarios. Ecological Engineering 65, 122-130.

Inagaki, S., 2000. Effects of proposed San Francisco Airport runway extension on hydrodynamics and sediment transport in South San Francisco Bay, Civil and Evironmental Engineering. Stanford University, Stanford, CA, p. 98.

Jackson, J.B., Kirby, M.X., Berger, W.H., Bjorndal, K.A., Botsford, L.W., Bourque, B.J., Bradbury, R.H., Cooke, R., Erlandson, J., Estes, J.A., 2001. Historical overfishing and the recent collapse of coastal ecosystems. Science 293, 629-637.

Jaffe, B.E., Smith, R., Torresan, L., 1998. Sedimentation and Bathymetric Change in San Pablo Bay: 1856–1983. USGS Menlo PArk.

Jaffe, B., Foxgrover, A., 2006. A History of Intertidal Flat Area in South San Francisco Bay,California: 1858 to 2005 U.S. Department of the Interior, U.S. Geological Survey, p. 32.

Jaffe, B.E., Smith, R.E., Foxgrover, A.C., 2007. Anthropogenic influence on sedimentation and intertidal mudflat change in San Pablo Bay, California: 1856–1983. Estuarine, Coastal and Shelf Science 73, 175-187.

Janauer, G.A., 2000. Ecohydrology: fusing concepts and scales. Ecological Engineering 16, 9-16.

Jarrett, J.T., 1976. Tidal Prism - Inlet Area Relationships., p. 59.

Jassby, A.D., Cloern, J.E., Powell, M.A., 1993. Organic carbon sources and sinks in San Francisco Bay: Variability induced by river flow. Marine Ecology Progress Series 95, 15.

Jassby, A.D., Cloern, J.E., Cole, B.E., 2002. Annual primary production: Patterns and mechanisms of change in a nutrient-rich tidal ecosystem. Limnology and Oceanography 47, 698-712.

Jolliff, J.K., Kindle, J.C., Shulman, I., Penta, B., Friedrichs, M.A.M., Helber, R., Arnone, R.A., 2009. Summary diagrams for coupled hydrodynamic-ecosystem model skill assessment. Journal of Marine Systems 76, 64-82.

Karunarathna, H., Reeve, D., Spivack, M., 2008. Long-term morphodynamic evolution of estuaries: an inverse problem. Estuarine, Coastal and Shelf Science 77, 385-395.

Kernkamp, H.W.J., Van Dam, A., Stelling, G.S., De Goede, E.D., 2010. Efficient scheme for the shallow water equations on unstructured grids with application to the Continental Shelf, Ocean Dynamics, p. 29.

Kimmerer, W., 2004. Open Water Processes of the San Francisco Estuary: From Physical Forcing to Biological Responses. San Francisco Estuary and Watershed Science 2.

Kineke, G.C., Sternberg, R.W., 1992. Measurements of high concentration suspended sediments using the optical backscatterance sensor. Marine Geology 108, 253-258.

Kirwan, M.L., Guntenspergen, G.R., D'Alpaos, A., Morris, J.T., Mudd, S.M., Temmerman, S., 2010. Limits on the adaptability of coastal marshes to rising sea level. Geophysical Research Letters 37.

Knowles, N., Cayan, D.R., 2002. Potential effects of global warming on the Sacramento/San Joaquin watershed and the San Francisco estuary. Geophysical Research Letters 29, 38-31-38-34.

Krone, R.B., 1962. Flume studies of the transport of sediment in estuarial shoaling processes. University of California, Berkeley, California

Latteux, B., 1995. Techniques for long-term morphological simulation under tidal action. Marine Geology 126, 129-141.

Le Hir, P., Cayocca, F., Waeles, B., 2011. Dynamics of sand and mud mixtures: a multiprocess-based modelling strategy. Continental Shelf Research 31, S135-S149.

Lesser, G.R., Roelvink, J.A., van Kester, J.A.T.M., Stelling, G.S., 2004. Development and validation of a three-dimensional morphological model. Coastal Engineering 51, 883-915.

Los, F.J., Blaas, M., 2010. Complexity, accuracy and practical applicability of different biogeochemical model versions. Journal of Marine Systems 81, 44-74.

Ludwig, F.L., Sinton, D., 2000. Evaluating an Objective Wind Analysis Technique with a Long Record of Routinely Collected Data. Journal of Applied Meteorology 39, 335-348.

Ludwig, K.A., Hanes, D.M., 1990. A laboratory evaluation of optical backscatterance suspended solids sensors exposed to sand-mud mixtures. Marine Geology 94, 173-179.

MacBroom, J.G., 2000. Tidal marsh restoration, OCEANS 2000 MTS/IEEE Conference and Exhibition. IEEE, pp. 1925-1928.

MacWilliams, M.L., Bever, A.J., Gross, E.S., Ketefian, G.S., Kimmerer, W.J., 2015. Three-Dimensional Modeling of Hydrodynamics and Salinity in the San Francisco Estuary: An Evaluation of Model Accuracy, X2, and the Low–Salinity Zone. San Francisco Estuary and Watershed Science 13.

Manh, N.V., Dung, N.V., Hung, N.N., Merz, B., Apel, H., 2014. Large-scale quantification of suspended sediment transport and deposition in the Mekong Delta. Hydrol. Earth Syst. Sci. Discuss. 11, 4311-4363.

Manning, A.J., Schoellhamer, D.H., 2013. Factors controlling floc settling velocity along a longitudinal estuarine transect. Marine Geology 345, 266-280.

Marciano, R., Wang, Z.B., Hibma, A., de Vriend, H.J., Defina, A., 2005. Modeling of channel patterns in short tidal basins. Journal of Geophysical Research: Earth Surface 110.

Marvin-DiPasquale, M., Cox, M.H., 2007a. Legacy Mercury in Alviso Slough, South San Francisco Bay, California: Concentration, Speciation and Mobility. U.S. Department on the Interior U.S. Geological Survey.

Marvin-DiPasquale, M., cox, M.H., 2007b. Legacy Mercury in Alviso Slough, South San Francisco Bay, California: Concentration, Speciation and Mobility, Open File Report. U.S. Geological Survey, Menlo Park, p. 98.

McDonald, E., Cheng, R., 1996. A numerical model of sediment transport applied to San Francisco Bay, CA. Journal of Marine Environmental Engineering 4, 1-41.

McKee, L.J., Ganju, N.K., Schoellhamer, D.H., 2006. Estimates of suspended sediment entering San Francisco Bay from the Sacramento and San Joaquin Delta, San Francisco Bay, California. Journal of Hydrology 323, 335-352.

McKee, L.J., Lewicki, M., Schoellhamer, D.H., Ganju, N.K., 2013. Comparison of sediment supply to San Francisco Bay from watersheds draining the Bay Area and the Central Valley of California. Marine Geology 345, 47-62.

Meshkova, L.V., Carling, P.A., 2012. The geomorphological characteristics of the Mekong River in northern Cambodia: A mixed bedrock–alluvial multi-channel network. Geomorphology 147–148, 2-17.

Milliman, J.D., Syvitski, J.P.M., 1992. Geomorphic/Tectonic Control of Sediment Discharge to the Ocean:

The Importance of Small Mountainous Rivers1 The Journal of Geology 100, 20.

Moffatt & Nichol Engineers, 2005. Hydrodynamic modeling tools and techniques,

South Bay Salt Pond Restoration Project. California State Coastal Conservancy.

Montalto, F.A., Steenhuis, T.S., 2004. The link between hydrology and restoration of tidal marshes in the New York/New Jersey estuary. Wetlands 24, 414-425.

Morgan-King, T., Schoellhamer, D., 2013. Suspended-Sediment Flux and Retention in a Backwater Tidal Slough Complex near the Landward Boundary of an Estuary. Estuaries and Coasts 36, 300-318.

Morris, J.T., Sundareshwar, P.V., Nietch, C.T., Kjerfve, B., Cahoon, D.R., 2002. RESPONSES OF COASTAL WETLANDS TO RISING SEA LEVEL. Ecology 83, 2869-2877.

Moskalski, S.M., Torres, R., Bizimis, M., Goni, M., Bergamaschi, B., Fleck, J., 2013. Low-tide rainfall effects on metal content of suspended sediment in the Sacramento-San Joaquin Delta. Continental Shelf Research 56, 39-55.

Nanson, G.C., Knighton, A.D., 1996. ANABRANCHING RIVERS: THEIR CAUSE, CHARACTER AND CLASSIFICATION. Earth Surface Processes and Landforms 21, 217-239.

Nicholls, R.J., Cazenave, A., 2010. Sea-Level Rise and Its Impact on Coastal Zones. Science 328, 1517-1520.

Niedoroda, A.W., Reed, C.W., Swift, D.J., Arato, H., Hoyanagi, K., 1995. Modeling shore-normal large-scale coastal evolution. Marine Geology 126, 181-199.

Obermann, M., Rosenwinkel, K.-H., Tournoud, M.-G., 2009. Investigation of first flushes in a medium-sized mediterranean catchment. Journal of Hydrology 373, 405-415.

Pasternack, G.B., Brush, G.S., Hilgartner, W.B., 2001. Impact of historic land-use change on sediment delivery to a Chesapeake Bay subestuarine delta. Earth Surface Processes and Landforms 26, 409-427.

Phillips, J.D., 2014. Anastamosing channels in the lower Neches River valley, Texas. Earth Surface Processes and Landforms 39, 1888-1899.

Postma, H., 1961. Suspended matter and Secchi disc visibility in coastal waters. Netherlands Journal of Sea Research 1, 359-390.

Powell, M.A., Thieke, R.J., Mehta, A.J., 2006. Morphodynamic relationships for ebb and flood delta volumes at Florida's tidal entrances. Ocean Dynamics 56, 295-307.

Prescott, K.L., Tsanis, I.K., 1997. Mass balance modelling and wetland restoration. Ecological Engineering 9, 1-18.

Ralston, D.K., Stacey, M.T., 2007. Tidal and meteorological forcing of sediment transport in tributary mudflat channels. Continental Shelf Research 27, 1510-1527.

Reed, D., 2002. Sea-level rise and coastal marsh sustainability: geological and ecological factors in the Mississippi delta plain. Geomorphology 48, 10.

Ridderinkhof, H., van der Ham, R., van der Lee, W., 2000. Temporal variations in concentration and transport of suspended sediments in a channel–flat system in the Ems-Dollard estuary. Continental Shelf Research 20, 1479-1493.

Rijn, L.C., 2011. Analytical and numerical analysis of tides and salinities in estuaries; part I: tidal wave propagation in convergent estuaries. Ocean Dynamics 61, 1719-1741.

Roelvink, J.A., 2006. Coastal morphodynamic evolution techniques. Coastal Engineering 53, 277-287.

Sampath, D.M.R., Boski, T., Loureiro, C., Sousa, C., 2015. Modelling of estuarine response to sea-level rise during the Holocene: Application to the Guadiana Estuary–SW Iberia. Geomorphology 232, 47-64.

Savenije, H.H.G., 2001. A simple analytical expression to describe tidal damping or amplification. Journal of Hydrology 243, 205-215.

Savenije, H.H.G., 2015. Prediction in ungauged estuaries: An integrated theory. Water Resources Research 51, 2464-2476.

Scavia, D., Field, J., Boesch, D., Buddemeier, R., Burkett, V., Cayan, D., Fogarty, M., Harwell, M., Howarth, R., Mason, C., Reed, D., Royer, T., Sallenger, A., Titus, J., 2002. Climate change impacts on U.S. Coastal and Marine Ecosystems. Estuaries 25, 149-164.

Scherrer, P., 2006. Le Havre – Port 2000: a move towards the environmental restoration of the Seine Estuary: Part 1.

Schoellhamer, D., 2011. Sudden Clearing of Estuarine Waters upon Crossing the Threshold from Transport to Supply Regulation of Sediment Transport as an Erodible Sediment Pool is Depleted: San Francisco Bay, 1999. Estuaries and Coasts 34, 885-899.

Schoellhamer, D.H., 2002. Variability of suspended-sediment concentration at tidal to annual time scales in San Francisco Bay, USA. Continental Shelf Research 22, 1857-1866.

Schoellhamer, D.H., Wright, S.A., Drexler, J., 2012. A Conceptual Model of Sedimentation in the Sacramento–San Joaquin Delta. San Francisco Estuary and Watershed Science 10.

Schoellhamer, D.H., Wright, S.A., Drexler, J.Z., 2013. Adjustment of the San Francisco estuary and watershed to decreasing sediment supply in the 20th century. Marine Geology 345, 63-71.

Shellenbarger, G., Downing-Kunz, M., Schoellhamer, D., 2015. Suspended-sediment dynamics in the tidal reach of a San Francisco Bay tributary. Ocean Dynamics 65, 1477-1488.

Shellenbarger, G.G., Schoellhamer, D.H., Lionberger, M.A., 2004. A South San Francisco Bay Sediment Budget: Wetland Restoration and Potential Effects on Phytoplankton Blooms, ASLO-TOS 2004 Ocean Research Conference, Honolulu.

Stive, M., Ji, L., Brouwer, R.L., van de Kreeke, C., Ranasinghe, R., 2011. EMPIRICAL RELATIONSHIP BETWEEN INLET CROSS-SECTIONAL AREA AND TIDAL PRISM: A RE-EVALUATION. 2011.

Stocker, T.F., Qin, D., Plattner, G.-K., Alexander, L.V., Allen, S.K., Bindoff, N.L., Bréon, F.-M., Church, J.A., Cubasch, U., Emori, S., Forster, P., Friedlingstein, P., Gillett, N., Gregory, J.M., Hartmann, D.L., Jansen, E., Kirtman, B., Knutti, R., Krishna Kumar, K., Lemke, P., Marotzke, J., Masson-Delmotte, V., Meehl, G.A., Mokhov, I.I., Piao, S., Ramaswamy, V., Randall, D., Rhein, M., Rojas, M., Sabine, C., Shindell, D., Talley, L.D., Vaughan, D.G., Xie, S.-P., 2013. Technical Summary, in: Stocker, T.F., Qin, D., Plattner, G.-K., Tignor, M., Allen, S.K., Boschung, J., Nauels, A., Xia, Y., Bex, V., Midgley, P.M. (Eds.), Climate Change 2013: The Physical Science Basis. Contribution of Working Group I to the Fifth Assessment Report of the Intergovernmental Panel on Climate Change. Cambridge University Press, Cambridge, United Kingdom and New York, NY, USA, pp. 33–115.

Suddeth, R.J., 2011. Policy Implications of Permanently Flooded Islands in the Sacramento–San Joaquin Delta. San Francisco Estuary and Watershed Science 9.

Sutherland, T.F., Lane, P.M., Amos, C.L., Downing, J., 2000. The calibration of optical backscatter sensors for suspended sediment of varying darkness levels. Marine Geology 162, 587-597.

Syvitski, J.P., 2007. Principles, methods and application of particle size analysis. Cambridge University Press.

Syvitski, J.P.M., Kettner, A.J., 2011. Sediment flux and the Anthropocene. Philosophical Transactions of the Royal Society A 369, 18.

Syvitski, J.P.M., Vörösmarty, C.J., Kettner, A.J., Green, P., 2005. Impact of Humans on the Flux of Terrestrial Sediment to the Global Coastal Ocean. Science 308, 376-380.

Takekawa, J., Woo, I., Athearn, N., Demers, S., Gardiner, R., Perry, W., Ganju, N., Shellenbarger, G., Schoellhamer, D., 2010. Measuring sediment accretion in early tidal marsh restoration. Wetlands Ecology and Management 18, 297-305.

Talke, S.A., Stacey, M., 2003. The influence of oceanic swell on flows over an estuarine intertidal mudflat in San Francisco Bay. Estuarine, Coastal and Shelf Science 58, 541-554.

Townend, I., 2012. The estimation of estuary dimensions using a simplified form model and the exogenous controls. Earth Surface Processes and Landforms 37, 1573-1583.

Townend, I., Fletcher, C., Knappen, M., Rossington, K., 2011. A review of salt marsh dynamics. Water and Environment Journal 25, 477-488.

Townshead, A., 2013. Vortex performs maintenance in Sacramento and Stockton Ship Channels, International Dreding Review. The Waterways Journal Inc., Missouri, USA.

van de Kreeke, J., 1985. Stability of tidal inlets—Pass Cavallo, Texas. Estuarine, Coastal and Shelf Science 21, 33-43.

Van de Kreeke, J., Robaczewska, K., 1993. Tide-induced residual transport of coarse sediment; Application to the EMS estuary. Netherlands Journal of Sea Research 31, 209-220.

van der Wegen, M., Roelvink, J.A., 2008. Long-term morphodynamic evolution of a tidal embayment using a two-dimensional, process-based model. Journal of Geophysical Research 113.

van der Wegen, M., Jaffe, B.E., Roelvink, J.A., 2011. Process-based, morphodynamic hindcast of decadal deposition patterns in San Pablo Bay, California, 1856–1887. Journal of Geophysical Research: Earth Surface 116, F02008.

van der Wegen, M., Roelvink, J.A., 2012. Reproduction of estuarine bathymetry by means of a process-based model: Western Scheldt case study, the Netherlands. Geomorphology 179, 152-167.

van der Wegen, M., 2013. Numerical modeling of the impact of sea level rise on tidal basin morphodynamics. Journal of Geophysical Research: Earth Surface 118, 447-460.

Van Ledden, M., Van Kesteren, W., Winterwerp, J., 2004. A conceptual framework for the erosion behaviour of sand–mud mixtures. Continental Shelf Research 24, 1-11.

van Maren, D.S., van Kessel, T., Cronin, K., Sittoni, L., 2015. The impact of channel deepening and dredging on estuarine sediment concentration. Continental Shelf Research 95, 1-14.

Vörösmarty, C.J., Meybeck, M., Fekete, B., Sharma, K., Green, P., Syvitski, J.P.M., 2003. Anthropogenic sediment retention: major global impact from registered river impoundments. Global and Planetary Change 39, 169-190.

Walling, D.E., Fang, D., 2003. Recent trends in the suspended sediment loads of the world's rivers. Global and Planetary Change 39, 111-126.

Wang, Y., 1998. Mixed effects smoothing spline analysis of variance. Journal of the royal statistical society: Series b (statistical methodology) 60, 159-174.

Wang, Z., Van Maren, D., Ding, P., Yang, S., Van Prooijen, B., De Vet, P., Winterwerp, J., De Vriend, H., Stive, M., He, Q., 2015. Human impacts on morphodynamic thresholds in estuarine systems. Continental Shelf Research, CSR3681.

Werner, B., 2003. Modeling landforms as self-organized, hierarchical dynamical systems. Prediction in geomorphology, 133-150.

Whipple, A., Grossinger, R., Rankin, D., Stanford, B., Askevold, R., 2012. Sacramento-San Joaquin Delta Historical ecology investigation: Exploring Patterns and Process, in: ecology, S.-A.s.h., Program (Eds.), California department of fish and game and ecosystem restoration Program. San Francisco Estuary Institute-aquatic science Center, richmond, Ca.

Whitcraft, C.R., Levin, L.A., 2007. REGULATION OF BENTHIC ALGAL AND ANIMAL COMMUNITIES BY SALT MARSH PLANTS: IMPACT OF SHADING. Ecology 88, 904-917.

Whitney, M., Jia, Y., McManus, P., Kunz, C., 2014. Sill effects on physical dynamics in eastern Long Island Sound. Ocean Dynamics 64, 443-458.

Williams, P., Faber, P.M., 2004. Design guidelines for tidal wetland restoration in San Francisco bay. Prepared for The Bay Institu.

Willmott, C.J., 1981. ON THE VALIDATION OF MODELS. Physical Geography 2, 184-194.

Winterwerp, J.C., Van Kesteren, W.G., 2004. Introduction to the physics of cohesive sediment dynamics in the marine environment. Elsevier.

Winterwerp, J.C., Manning, A.J., Martens, C., de Mulder, T., Vanlede, J., 2006. A heuristic formula for turbulence-induced flocculation of cohesive sediment. Estuarine, Coastal and Shelf Science 68, 195-207.

Winterwerp, J., Wang, Z., 2013. Man-induced regime shifts in small estuaries—I: theory. Ocean Dynamics 63, 1279-1292.

Wright, S.A., Schoellhamer, D.H., 2004. Trends in the Sediment Yield of the Sacramento River, California, 1957 - 2001. San Francisco Estuary and Watershed Science 2.

Wright, S.A., Schoellhamer, D.H., 2005. Estimating sediment budgets at the interface between rivers and estuaries with application to the Sacramento-San Joaquin River Delta. Water Resources Research 41, W09428.

Yahg, S.L., 1998. The Role of Scirpus Marsh in Attenuation of Hydrodynamics and Retention of Fine Sediment in the Yangtze Estuary. Estuarine, Coastal and Shelf Science 47, 7.

Yang, Z., Wang, T., 2012. Hydrodynamic modeling analysis of tidal wetland restoration in Snohomish River, Washington, 12th International Conference on Estuarine and Coastal Modeling 2011, St. Augustine, FL, pp. 139-155.

Exposure

Journal papers

Achete, F. M.; van der Wegen, M.; Roelvink, D., Jaffe, B. :   A 2D Process-Based Model for Suspended Sediment Dynamics: a first Step towards Ecological Modeling, Hydrol. Earth Syst. Sci., 19, 2837–2857, 2015 doi:10.5194/hess-19-2837-2015

Achete, F. M.; van der Wegen, M.; Roelvink, D., Jaffe, B.:   Suspended Sediment Dynamics in a tidal channel network under Peak River Flow, Ocean Dynamics DOI: 10.1007/s10236-016-0944-0

Achete, F. M.; Reys, C.V., van der Wegen, M.; Roelvink, D., Foxgrover, A., Jaffe, B.: Impact of a sudden tidal prism increase in estuarine sediment flux: implications to remobilization of Hg-contaminated sediment. Science of Total Environment (submitted)

Achete, F. M.; van der Wegen, M., Jaffe, B.; Roelvink, D.:   How important are climate change and foreseen engineering measures on the sediment dynamics in the San Francisco Bay-Delta system? Environmental Science and Technology.(submitted)

Conference presentations and posters

Achete F.M, van der Wegen, M.; Roelvink, D., Jaffe, B.: Assessing Suspended Sediment Dynamics in the San Francisco Bay-Delta System: Coupling Landsat Satellite Imagery, in situ Data and a Numerical Model. 16th The Physics of Estuaries and Coastal Seas (PECS) Symposium, New York, United States of America, August-2012 [poster presentation].

Achete F.M, van der Wegen, M.; Roelvink, D., Jaffe, B.: Tracking sediments through the Bay-Delta system over a water year with a 2D process based model (D-FLOW FM). 7th Biennial Bay-Delta Science Conference, Sacramento, United State of America. October-2012. [oral presentation].

Achete F.M, van der Wegen, M.; Roelvink, D., Jaffe, B.: Assessment of long-term impacts due to SSC changes in San Francisco Bay-Delta. American Geophysical Union (AGU) Fall Meeting, San Francisco, United States of America December-2012, [poster presentation].

Achete F.M, van der Wegen, M.; Roelvink, D., Jaffe, B.: Sediment Transport in a Complex Estuarine Channel Network. 22rd Biennial Coastal and Estuarine Research Federation (CERF) Conference. San Diego, United States of America, November-2013 [oral presentation]

Achete F.M, van der Wegen, M.; Roelvink, D., Jaffe, B.: 2D Process-Based Model for Assessment of Suspended Sediment Budget. NCK Days. Delft, The Netherlands, March-2014 [oral presentation]

Achete F.M, van der Wegen, M.; Roelvink, D., Jaffe, B.: Automated Calibration of Suspended Sediment Concentration Levels for Process-based Modeling in Estuaries. 17th The Physics of

Estuaries and Coastal Seas (PECS) Symposium, Porto de Galinhas, Brazil, October-2014 [oral presentation].

Achete F.M, van der Wegen, M.; Roelvink, D., Jaffe, B.: Sediment trapping efficiency in the San Francisco Bay-Delta system by means of a 2D process-based model. 23rd Biennial Coastal and Estuarine Research Federation (CERF) Conference. Portland, United States of America, November-2015 [oral presentation]

Biography

Fernanda Minikowski Achete was born in Rio de Janeiro, Brazil, on 6 September 1986. . She obtained her B.Sc. in Oceanography, in 2008 from the Faculty of Oceanography, State University in Rio de Janeiro, Brazil. While studying, she was already working research projects at Federal University of Rio de Janeiro and after the graduation she became junior researcher. Her main responsibilities included modeling ocean circulation at the Atlantic Ocean and field work. The Piatam Oceano and REMO projects were a partnership between the Brazilian oil company Petrobás, the Brazilian navy and the University.

In 2011, she obtained her MSc in Coastal Engineering and Maritime Management (CoMEM) from an Erasmus Mundus partnership between Technical University Delft (TU Delft, the host university), Delft The Netherlands; Norwegian University of Science and Technology (NTNU), Trondheim, Norway and Polytechnic University of Catalonia, Barcelona, Spain. The thesis topic was beach morphodynamics, investigating the evolution of the mega nourishment the "San Engine". Her thesis entitled "Ameland Bornrif: a case study for the sand engine" was funded by Deltares as part of the project "Building with the Nature".

In October 2011, she started working on her PhD research at the Department of Water Science and Engineering at UNESCO-IHE/TU Delft. Her research was part of a larger project Computational Assessments of Scenarios of Change for Delta Ecosystem (CASCaDE II), a United State Geological Survey (USGS) multidisciplinary project funded by Calfed. In addition to the 3 USGS offices, the project involved 3 universities and Deltares. This research was also partially funded by the Brazilian Government via Capes agency.

Netherlands Research School for the
Socio-Economic and Natural Sciences of the Environment

D I P L O M A

For specialised PhD training

The Netherlands Research School for the
Socio-Economic and Natural Sciences of the Environment
(SENSE) declares that

Fernanda Minikowski Achete

born on 6 September 1986 in Rio de Janeiro, Brazil

has successfully fulfilled all requirements of the
Educational Programme of SENSE.

Delft, 12 April 2016

the Chairman of the SENSE board   the SENSE Director of Education

Prof. dr. Huub Rijnaarts   Dr. Ad van Dommelen

The SENSE Research School has been accredited by the Royal Netherlands Academy of Arts and Sciences (KNAW)

KONINKLIJKE NEDERLANDSE
AKADEMIE VAN WETENSCHAPPEN

The SENSE Research School declares that Ms Fernanda Minikowski Achete has successfully
fulfilled all requirements of the Educational PhD Programme of SENSE with a
work load of 39.2 EC, including the following activities:

SENSE PhD Courses

- Environmental Research in Context (2013)
- Research in Context Activity: 'Co-organising the SENSE writing week' (2013)
- SENSE writing week (2013)
- SENSE Summer Academy 'Emerging Issues in Sustainable Energy and Water Systems'
  (2013)

Other PhD and Advanced MSc Courses

- Sediment Dynamics, Delft University of Technology (2014)
- Summer School 'Estuarine and Coastal Process in relation to Coastal Zone Management',
  Netherlands Centre for Coastal Research (NCK) (2014)

External training at a foreign research institute

- Training in ADV (for sediment concentration), calibration of ADV , ADCP (for currents),
  and CTD equipment, United States Geological Survey (USGS) (2013)
- Collaboration in the CASCaDE project, developing the DFlow FM model and working with
  the DELWAQ models on hydrodynamics and sediment dynamics, United States
  Geological Survey (USGS) (2012, 2013)

Management and Didactic Skills Training

- Supervising two MSc students with thesis entitled 'Modelling sedimentation in San
  Francisco Estuary' (2011) and 'Mud Dynamics in a tidal channel: impact of opening salt
  ponds on channel deepening' (2014), UNESCO-IHE Delft
- Member of the PhD association board (PhD council), UNESCO-IHE Delft (2012-2014)

Oral Presentations

- *Tracking sediments through the Bay-Delta system over a water year with a 2D process
  based model (D-FLOW FM)*. 7th Biennial Bay-Delta Science Conference, 12-13 October
  2012, Sacramento, United State of America
- *Sediment Transport in a Complex Estuarine Channel Network*. 22nd Biennial Coastal and
  Estuarine Research Federation (CERF) Conference, 1-6 November 2013, San Diego,
  United States

SENSE Coordinator PhD Education

Dr. ing. Monique Gulickx

T - #0431 - 101024 - C156 - 244/170/8 - PB - 9781138029767 - Gloss Lamination